QA
95
.H64
1973

DATE DUE

MAR 1 8
FEB 0 8 '93
DEC 0 6 '93
DEC 1 3 1996
DEC 1 9 1997 S

WITHDRAWN

GAYLORD 234
PRINTED IN U.S.A.

Let's Play Math

MICHAEL HOLT and ZOLTAN DIENES

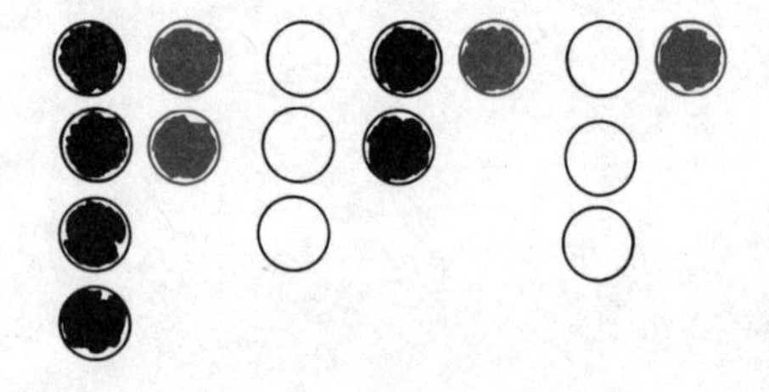

Let's Play Math

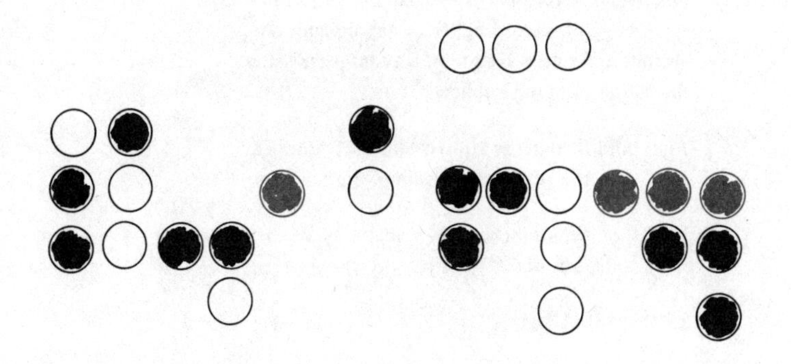

WALKER AND COMPANY
New York

Copyright © 1973 by Michael Holt and
Zoltan Dienes.

All rights reserved. No part of this book may
be reproduced or transmitted in any form or by
any means, electronic or mechanical, including
photocopying, recording, or by any information
storage and retrieval system, without permission
in writing from the Publisher.

First published in the United States of America
in 1973 by the Walker Publishing Company, Inc.

Published simultaneously in Canada by Fitzhenry &
Whiteside, Limited, Toronto.

ISBN: 0-8027-0418-2

Library of Congress Catalog Card Number:
72-95791

Printed in the United States of America from
type set in the United Kingdom.

10 9 8 7 6 5 4

Contents

260541

To
Miranda and Paul

The Playwiths Kit

Playwiths

Sticks 'n Stones
Shop and Swop
Sesame Dice
Rainbow Toy
Shape Toy
Triangle Toy

Card Games

Secrets – with 12 cards
Code cards – with 6 cards
Passtime – with 8 cards
Happy Sets – with 24 cards
Blobs – with 16 cards

PART 1

Math Can be Child's Play

CONSTRUCTION
LEONARDO

Thinking for tomorrow

It is a truism to say that math is useful. Most business men, accountants, engineers, mechanics, farmers and scientists will bear this out. But what many mean by math – numbers and calculation – is only part of the story. Math is also about such unlikely things as family relationships (brother-sister ties, for instance), puzzles about drawing shapes without lifting your pencil, and paper cutting. These are what make the new math such fun for children to do and, incredibly, so vital if they are to flourish effectively in tomorrow's newly hatching technological world. No longer is it enough to train children to meet the known challenges: the generation gap, over-population, mass urbanization of the world, automated society, computerized business, and the shrinking of our world to a McLuhan-sized 'global village' through the impact of the new mass media – reason enough to rethink our approach to math learning; they must be helped to face the unknown, for one thing seems certain, whatever tomorrow is like it won't be much like today.

How are we to educate children to meet the looming, unpredictable challenges of the year A.D. 2000? One way, many educators believe, is through the presentation of novel and strange problems in the form of activities and thinking games, the general flavour of which may be glimpsed in the photo.

Child psychologists have been astonished to find how much a very young child is capable of learning. If he can learn to speak, single-handed and without a teacher, by the time he is two years, no wonder he can learn so much more than was once thought possible. The secret lies in catching him young enough before his mind 'hardens'. Experience shows that youngsters, if started young enough, can learn such seemingly 'way-out' things as logical reasoning, mathematical patterns and how words are put together. It is such ideas that are built into the structure of our games.

Now we are not suggesting our games will really *teach* a child math. But classroom experience indicates they encourage an alert, open-minded attitude in youngsters and help them develop their potential for clear thinking. Even more important – the games are fun to play.

Tomorrow's math today

The new math is not a passing phase which, if, like a James Thurber character, you look the other way, will go away again. It is here for good. The reason: trad. math, especially as it used to be taught, was largely about the tricks of the mathematical trade; the new math, on the other hand, reveals the underlying laws that govern those tricks. By the year A.D. 2000, when every well-to-do household will sport an on-line terminal to a computer, we shan't need people to slog through routine 'sums'; computers will do that. But we *shall* urgently need people – men and women – to prepare the computer's programs – to do its *thinking*. For this reason alone, the new math has become more interesting and needful. And, with luck, it will become more so.

Now this quiet revolution in math has brought in its train another equally exciting revolution – in the way we teach youngsters to think, do math, and read.

Research over the past twenty years has made it possible at least to sketch in the basic principles by which learning, particularly of math, can be made easier and more fun for the clever and ordinary alike. And even for the not very gifted child, too. There has been a growing conviction that youngsters learn best through activities and games. It was in the light of this research – some of it conducted by one of us, Zoltan Dienes – that we carefully devised our games.

The games

If math can be turned into a game, it can literally become child's play. This ideal in mind, we invented more than eighty new games, presented here. Each game was devised to develop some aspect of a child's inborn potential to think creatively; most, but not all, have a mathematical flavour; some have even found use in developing artistic flair. They are for the parent or teacher to play constructively with youngsters of four years and up; some are even suitable for seven-year-olds. Most of the games can be played in the home or kindergarten with little adult supervision, once the rules have been learnt. We have deliberately devised novel and puzzling play situations. For children and adults alike are enormously attracted to puzzles. Also, we believe, children relish a certain amount of the uncertain.

The games conceal interesting mathematical patterns. These the child may discover for himself. If he does, he will have acquired a taste of what math is really about; if he doesn't, no matter: he will have enjoyed himself. It is a sad fact that few children in the past – who may be reading this now – derived much enjoyment, let alone technical mastery, from their math lessons. All the games, it must be emphasized, are simple and brief. We do not expect parents to know how to keep a game going as trained teachers do.

The playwiths

We have given a special name to the cards and toys with which many of the games are played: playwiths. This was quite deliberate. The playwiths are not simple toys; they conceal a mathematical pattern or structure. Young children's games sold in shops could be made much more mentally challenging – not, of course, that all games should be intellectually demanding! Because of the unusual mathematical ingredients in the playwiths they may seem strange to the adult. But research among many sub-cultures shows they belong comfortably in a child's fantasy world.

Six separate playwiths and five packs of cards are needed to accompany the games. All these can be made easily and cheaply by an adult out of paper, cardboard and scrap wood. The card faces can be drawn or traced; the playwiths constructed with scissors and paste; the woodblocks, used for the Number games (Set 9), can be sawn by the average handyman.

Why have playwiths?

You might possibly wonder whether it isn't cheating to make up special playwiths. Why not, you might say, let a child discover mathematics in his surroundings? A nice idea and one advanced by many educationists. The trouble is, math is not abundantly obvious in our surroundings. Admittedly, one may ask a child to find oblongs in a brick wall or triangles in the girders of a bridge. But that's like asking a child to find the sweets in a bag of sweets: the sorting has been mostly done already. With our playwiths the mathematical content – shape, color or size – is partially hidden. A degree of discrimination is needed to unearth it, just as in unravelling a good mystery or puzzle.

By no means are the playwiths intended to replace the many excellent educational pieces of 'structural' apparatus used in infant classrooms today. These include the multi-base blocks designed by Zoltan Dienes and tested over a decade or more in many countries, Caleb Cattegno's Cuisenaire rods and Catherine Stern's apparatus. All have a place in the classroom. The playwiths work in accordance with simple rules just like chess, draughts, and bridge. The rules are what give the games their edge, make them fun to play. For instance, games like draughts can be modified to have the same underlying patterns of rules as the turns and flips of a cube, as is shown on the board in the photograph overleaf.

Reading readiness

Our games were designed, as we said, to foster new ideas in the child's mind which he will naturally want to talk about and, later, record in writing. The part language plays in the formation of early concepts is now known to be crucial. Without the creative use of language a child's gift of intelligence cannot mature.

The games also induce a child to think with his hands. Research shows that most children see problems in terms of their own body. Mathematics in particular requires strong links to be made between hands and brain for success. The games go some way towards establishing a brain-hand link. Through such a link they promote hand skills and artistic expression.

All these activities are useful for building the child's four vocabularies: hearing, speaking, reading and writing.

Hearing and speaking are promoted by sand-and-water play and verbal games. Our games can be a novel addition to these staples of infant education.

Reading readiness is promoted by our simpler games such as sorting and matching by color, shape and size in various materials. Without such practice a child finds it hard to discriminate between shapes and, thus, later the letters of the alphabet. Active handling of our home-made playwiths develops a child's sense of touch. This in itself is a valuable stimulus to description. The card games involve reading labels. This fosters an ability to read letters.

Writing skills are encouraged through the making of pictures and patterns. These may be in paint, in sand, or in pencil.

The Games and the 'New' Math

IMPORTANT ADVICE! One thing we must emphasize: the mathematics in the games has a very special quality. In the first place it does not much resemble the usual picture of math—traditional or modern. Its roots are in the researches of certain creative mathematicians. In this sense it is more *avant garde* than the "new math" as taught in schools—"new math" in name only, for much of it was invented a century ago! In the second place, the math in the games is more abstract than even many teachers will allow. At first sight it seems "tough going" . . . but only, as it happens, for the adult. Not for the child. The games were specially designed with *children* in mind. So to get the best out of them (see Part 2) it is necessary to give them a fair trial. And don't expect them to resemble the standard "new math" now being taught in most schools. Such games as we describe may well be common among children in a generation or so.

The magic word 'set'

The link between these games and mathematics may not be obvious. To play them you need what appears to be rather unmathematical equipment – twigs, shells, bottle tops, buttons and the like. How can such mundane things help a child do math? The answer lies not in *what* the things are but in the way they are played with.

Even the preliminaries to a game can involve a child in thinking mathematically. Before he can use the playwiths he has to make distinctions about their appearance. That is, he must consider the *properties* of the things, as a mathematician would say.

Then he goes on to look for common properties among various other playwiths which are specially designed to make this easy. That done, he can lump several playwiths together with a common property – perhaps they are all red or all soft or all big. He puts them into a *set* – a key word in the new math.

Later still he can give the set a name or a label. For instance, all the sets with the same number of things in them as there are eyes in one's head go by the name of 'two'. From the name 'two' the child is led on to the idea of linking up two things and another two things: he sees how they relate; he builds up mathematical links, or *relations,* as they are known technically. The idea of a relation is so important in maths that we devote the whole of the next few sections to it.

The early games (Sets 1 and 2) on sorting develop a child's conception of 'property' and 'set'. Later games build on the early concept of a 'set' to establish early ideas of algebra (Sets 6, 7 and 10) and of number (Set 9).

Math is relating things

Math is not, as we said, all numbers. Numbers do, of course, play a vital role in a child's mathematical education, as they do in our games. It is actually the study of how things link up, relate (much as brothers and sisters or uncles and cousins do!) and, at a higher level, how these link-ups themselves relate.

What does a mathematician mean by 'relate'? A helpful illustration is provided by the way shirts are arranged (sorted) in a department store. The sales assistant has, we'll say, put all the bright pink shirts in one drawer. Each shirt in that drawer is related to every other shirt there simply by being the same color, pink. (Contrariwise, a mathematician would be equally happy to relate two shirts because one is pink and the other is not!) In another drawer the assistant pops all the blue shirts, in a third all the green ones. Within each drawer – technically, a *set* – each shirt is related by being the same color; but the drawerfuls themselves are related among themselves in quite another way. First, we related one shirt to another within a drawerful (our set); now we are relating one drawerful (set) to another (set). We are relating whole blocks of relations – the ones relating shirts to shirts, such as their all being red or all being blue, and so forth. We are relating, then, what we have already related. A distinct feat of mathematical abstraction on our part!

Several innocent-seeming relations – social, geographical, linguistic – can be rendered mathematical, as we shall now see. In other words a child can extract math from his surroundings but less straightforwardly than is usually envisaged.

Social relations

Depend as we do on these, social relations are not a feature of the games. This is not because they cannot be analysed mathematically – they can, albeit crudely, as we shall show: they simply do not enter a child's ego-centred world.

Take the case of Bill, an executive in a company. His boss, Ralph, is Vice-President. What is their relation to each other in the company? Ralph's to Bill's is the boss relation. But the relation is not two-way: Bill obviously is not Ralph's boss. The only situation in which such a paradoxical relation could hold would be in a Lewis Carroll world. If they were both each other's boss, the relation would be perfectly symmetrical, as the mathematician would say. An instance of such a paradoxical two-way relation occurs in the political joke:

What is capitalism?

The exploitation of man by man.

And communism?

The reverse.

The joke suggests the exploitation relation is a two-way relation. But, as everyone knows, it is unfairly one-way and highly unsymmetric.

But what possible bearing can this have on arithmetic? An older child than we are writing for will meet this two-way relation in numbers. He will know, for instance, that 5 is more than 3. The 'is more than' relation is not, however, a two-way one, for 3 is not more than 5. The properties of relations such as 'is less than', 'is more than', 'is equal to' have their counterparts in the operations we do with numbers – operations like multiplying or subtracting. By being presented with the properties of relations outside the context of number problems, a child may learn to see the wood of relations in the trees of number manipulations.

A more complex social relation is the order in which names are listed in a telephone directory. Although we offer nothing as complex as this, the Order games in Set 7 guide a child through his first steps to the order numbers – first, second, third and so on.

Geographical relations

Almost as important are geographical relations. They have an obvious bearing, too, on our ideas of space and geometry.

Suppose you go for a walk in a strange country. On the way out, so as to be able to find your way home again, you must gather *cues*, as psychologists would say – landmarks such as blasted oaks, fallen branches, cart tracks, boulders, streams and so forth, and forks and junctions in the path. These last can present you with problems – problems, as it happens, in relationships of a most taxing kind. On a walk you might easily overlook a track that joins yours from the right, say, from 'behind you' as it were. On the return journey, faced now with a fork, you cannot be sure whether to bear right (the way you came) or left (down the strange track). You might settle the matter by proceeding a short way along each path and deciding from the look of the trees and plants; or you might relate the turn of your tracks now to the way they turned on the way out – 'Before I veered left near this fork. So now I must walk in a right curve,' you think. Again you could relate your path to distant landmarks such as houses or even the sun. Because of the complexity of the relations involved, explorers solve the problem by carrying a compass. ('I turned East on the way out, so on the way back I veer West.') Next time you are out walking with your child, see whether he can find his way home without your telling him. You can turn the walk into a game.

So many schoolbook problems in geometry are couched in geographical terms that the value of such games is clear. More immediately in a child's school career, he will need to handle relations of 'opposite to' or 'the same as' for these ideas:

left and right
up and down
half-turn and quarter-turn
sideways and forwards
square and slanting
North and South
and so on.

These concepts are part and parcel of a child's vocabulary of space. We make use of them to develop his sense of space in the Maze and Dance games (Set 5). Although they are not difficult concepts to grasp they are probably the ones most taken for granted in a child's education; the ones it is assumed he will pick up as he goes along. It seems likely that the child from the enriched background will acquire them on the way; but less likely that children of impoverished Sesame Street backgrounds will be so lucky. One of the authors, Holt, has played several of the spatial games in Set 5 at his home as a weekly math party for local children, like that shown in the photograph: here the children are arranging a set of horses' heads on cards to fit a pattern of turns shown on the board by arrows.

He conducted a mathematical fun-fair stall with 'Sesame Street' kids at a Festival arranged by Joan Littlewood in London.

Relations in language

Today children learn sentences almost like math – as structured blocks: subject + verb + the rest. The Code card games in Set 3 reflect this trend. Take the simple sentence: 'The dog eats the bone.' The verb 'eats' expresses a relation which is certainly not two-way. If it were, the bone could eat the dog. The games symbolize for children the order of building up the units of a sentence, a vital aspect of reading comprehension.

In English, at any rate, we express such relations by the order of the words. In other languages, notably Latin, the order of the words can be changed without altering the expressed relationship. An interesting pastime can be to invert the order of adjacent words in each line of a well-known story or nursery rhyme. For instance, 'Mickey Mouse' becomes 'Mouse Mickey'. See if you can decode this:

> Mother Old went Hubbard the to cupboard
> fetch To poor her a doggie bone,
> she When there got, cupboard the bare was
> so And poor the had doggie none.

It is remarkable, when you think about it, how easily a child will decode this garbled German-sounding version. But if one were to upset the order of subject, object and verb – that is, to disguise the verb relationship, it would present a child with an insuperable problem. What, for instance, is the intended order in the garbled sentence: 'Slew Abel Cain', even knowing which is the verb, 'slew'?

Family ties

Improbably, even family relations can be viewed in a mathematical light. Glance for a moment at this ordinary family tree illustrating the result of the union of the timeless Joneses and Robinsons.

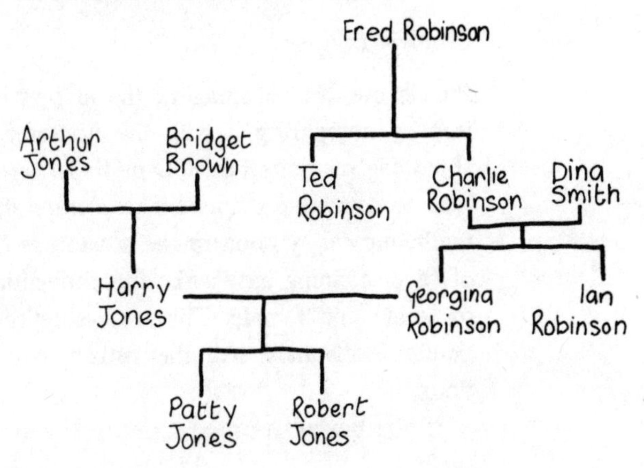

Then look at a mathematical version. The capital letters are the initials of the Christian names.

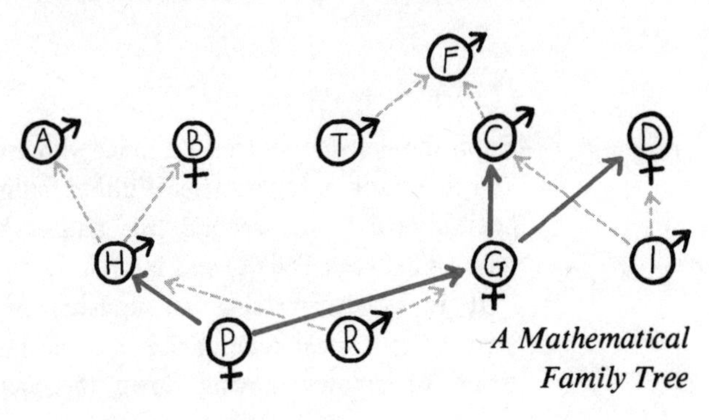

A Mathematical Family Tree

the people — the relations

♂ means male — ↑ (grey) means "is the son of"

♀ means female — ↑ (red) means "is the daughter of"

Problem 1

Puzzle out the meaning of the arrows in the second tree by comparing it with the first tree. The arrows stand for relations but not, perhaps, ones that would leap to everyone's mind! The picture shares with all mathematical symbolism the advantage for the reader of 'eye-scanning ease', like the conventional symbols for male and female. The arrows represent not particular individuals, like the initials, but general relations.

If you haven't worked it out, you can readily check that:

a grey arrow stands for the relation 'is the son of'
a red arrow for 'is the daughter of'

Problem 2

From the first family tree it is clear who are the uncles: Ted Robinson is Ian's uncle. Without finding any more uncles, turn to the second tree and trace the uncle relation between Ted (T) and Ian (I).

It is worth stressing an unusual aspect of the arrows: the uncle relationship can be made up of a string of arrows moving down the page, with the stream of time. (Note: the son and daughter arrows point *up* the page against the stream of time.)

The uncle relation can then be represented by the following string of arrows:

up a grey

down (against the arrow) either a red or grey

and down either a red or grey.

Check with I (Ian). From I move up a grey arrow to reach D; then down a red and a red to reach P. And I is indeed the uncle of P (Patty). From I move up the grey arrow to D again then down a red and a grey, to reach R. A glance at the first family tree confirms I to be the uncle of R (Robert) as well. Further, T (Ted) is also G's (Georgina's) uncle.

These two problems are intended to highlight something of the subtlety of thinking that can be involved even in apparently trivial problems such as tracing a family tree. Needless to say, the relation games in Set 6 are much simpler. Though we must emphasize that to a child the simple relations of 'more than' or 'less than' can be equally puzzling.

How can we translate our family tree arrow relations to numbers? In line with modern trends we draw a so-called number ladder or number line. Ours runs from 1 to 5:

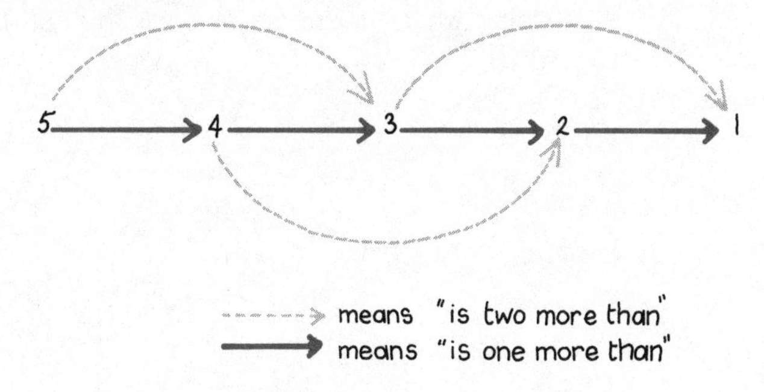

Fancifully, one might call the grey arrow the 'is a grandfather of' arrow, the red arrow the 'is the father of' arrow. Most arrow diagrams show the arrows going the other way: $3 \longrightarrow 5$, meaning 5 is 2 more than 3. We favour our presentation for sound mathematical reasons; the other presentation is a grammatical view. Whether the reader wishes to follow these technicalities or not he will at least appreciate that interesting subtleties exist.

What puzzles most children is that a couple of '1 more' arrows is in effect the same as a single '2 more' arrow. Granted, they might learn masses of particular number facts such as $3+2=5$, and so on; but that is no guarantee they grasp the general rule that two red arrows are the equal of one grey arrow on the number line. And without such an understanding they are in no position to cope with the generalities of algebra either.

The last set of games, Pebble games (Set 10) are basically a first primer on algebra – though we would implore the reader not to breathe a word to the children.

Our own experience shows that children raised on such ideas definitely tackle the standard bits of math – the four rules, tables, and so on – with more zest and understanding. If the truth be known, most of us like doing what we *can* do. And we do it all the better for that.

PART 2

How to Get the Best out of the Games

Child psychology

Whenever the playwright James Barrie had literary friends to tea, talk inevitably turned to the latest novel of the day . . . and Nana, his dog, would pad across to the bookcase, to retrieve there the very book they were discussing, to everybody's astonishment. Barrie had put the book he was going to talk about in a special spot in the bookcase and had trained Nana to pick it out. He was like the old-style teacher. Teachers once trained children to perform similar, mathematical tricks, which give some parents goose-flesh merely to think about them. The tricks are as tiresome to learn as, say, metres and metres of old-fashioned poetry. In the new math, the tricks take a back seat. It is understanding the reasons for them that becomes the main point of teaching math.

Now we are not suggesting that anyone can be a mathematician, any more than we can all be conversationalists, though we can all talk. The point we are making is that you can enjoy thinking through a new puzzle with your child. What these new math games aim to promote is enjoyment in thinking and, thereby, in math. This part of the book is offered as a crash-course on child psychology. It has the modest aim of helping parent and teacher alike to get the best out of these games with a child.

Get to know him

Most mothers may be inclined to think, 'I'm far too busy to play math with my child; after all, he'll pick it up at school later.' Many fathers get home, wanting nothing better than to sink into a chair and read the evening paper. But if parents would realize how valuable their interest can be at the early stages in their child's life, they would be more inclined to feel the value of playing with him. We don't mean that parents should try to give their child a flying start at school by cramming him. The benefit from your playing math games with him may not show up till much later. The benefit may, of course, be an unseen one.

But the very act of playing with a child is often enough to give him confidence. A little effort can pay big dividends. We stress 'little' because five or ten minutes of real play with him is worth half an hour of your going through the motions out of a sense of duty. That isn't going to fool him for a moment. 'Does she really like playing trains?' a child will wonder of his mother. 'How odd!'

Your approval has another side: research shows many delinquents have felt themselves cut off from their parent's approval, and so from society's. We don't want to give the idea that the parent who doesn't play math games with his child is heading for trouble. But people depend so much on television that they forget the enjoyment of making their own amusements.

Stage-manage mistakes!

Don't worry about making mistakes. To err is human. Even make deliberate mistakes, for two good reasons.

First, if you do, then it will not show when your demonstrations of games go wrong (as they are bound to!). Second, deliberate mistakes indicate whether a child has really grasped an idea or is simply going through the motions, as they say. You might try counting things badly. If you aren't corrected by the child, then something is severely amiss. Stage-managed mistakes can encourage a child to think for himself rather than obey the rules simply to please you.

Don't scold errors

Never make a moral issue of a child's mistakes. This is probably the most vital piece of advice we have to offer, and for two reasons. First, a child's slip can be both a useful part of his learning and a tell-tale sign for the alert teacher. If one attaches a moral significance to errors and records them by ticks and crosses, then the ticks and crosses can become symbols for 'goody-goody' and 'naughty'. When really they should be treated as useful evidence of learning on the child's part.

Second, when errors become a moral issue, a child will naturally prefer not to play – that is, not to learn – rather than risk being put in the wrong and being made a fool of. So never scold unsuccessful tries.

On the other hand, don't encourage him always to strive for the correct answer. If you do, he'll learn to please you and not to tackle the problem in hand. He'll be in danger of becoming 'cute'.

Keep things moving

This applies in both senses. All children learn best by moving and doing, which in turn trigger off thinking.

Anyway, to tell your child to sit still is like putting him in a strait-jacket. (You try learning something without moving a muscle!) And it's best to keep things moving in the other sense: change the game before it gets stale.

Don't expect too much

It is easy to forget how slow and clumsy a youngster can be. For example, watch a child fold a sheet of paper in two. It is misguided aid to make the fold neat and square for him. That's like the parent who always helps a child tie his laces. Far better to let him struggle. Otherwise he'll never learn.

The adult has a much more vital role to play than 'nursemaid' – that of an encouraging, knowledgeable friend.

Even the seemingly simple task of copying presents sizeable problems to a child. You draw for your child three identical 4-by-3 grids of squares, each on its own sheet of paper.

Fix two of the sheets to the wall; lay the third on the floor. Mark an 'X' in any square on one of the wall

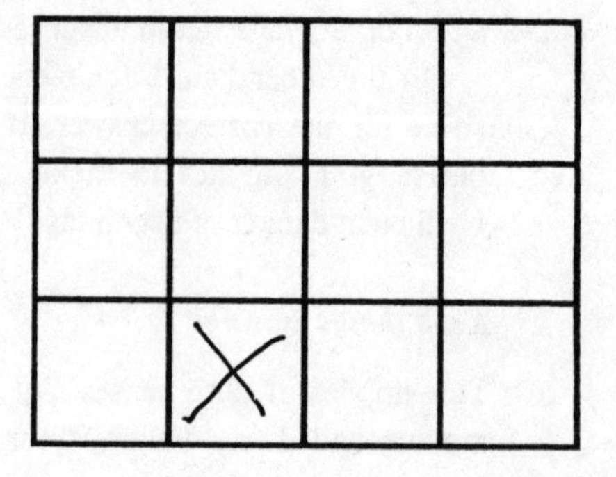

grids. Ask your child to mark the same, matching square on the other wall grid. He'll probably put an 'X' in the matching square without any trouble. A four-year-old, however, might find it hard to do. But ask him to put a cross in the matching square on the grid on the floor and, eleven to one, he'll mark the wrong square! Which just goes to show that adults mustn't expect overmuch of a child, however simple the task. It takes a pretty astute parent to know what a young child is capable of.

Take your time

When a child doesn't want to play with you, it is as a rule wiser to give in gracefully. Let him wander off to do what he wants. If you force him, he'll only wriggle out one way or another. (See in this part the section *Don't try to trap him.*)

Perhaps, a mother, who has gone to great lengths to prepare a game, understandably can't wait to get her boy started. 'But I'm going to paint,' he declares firmly. 'No, you're not,' she insists. 'We're going to play the Pebble game.' All he wants to do is mess about with paints. Much better she should humour his wish than impose her own. All the same, the parent *has* a subtle part to play in the job of interesting her child.

Don't let an over-eager child go on too long; try to keep the games down to half an hour, which is about all a youngster can take. Always stop before he gets bored; and if he wants more, suggest a different sort of game. There's a wide selection of games for you to try here. Variety is, after all, the spice of learning.

And always let him play at his own speed – especially when he's picking up a new game – or he'll never hit on the ideas behind the games. Then is the time to take the passenger seat *patiently* and avoid back-seat driving.

A child enjoys doing things in his own time. Research shows that playing around can be as serious to a child as work is to an adult. It just looks like messing about to a grown-up, that's all. Executives always say they get their best ideas when soaking in the bath or lolling in bed. So idling about can be recognized as valuable by adults.

Finally, you can't hurry a child. Nor can you force him. If he doesn't want to play a game, try another one or let him go his own way. That way you keep your patience and his affection.

One of the most satisfying moments in a young child's education is when the 'penny drops'. So avoid talking at your child as much as possible; let the penny drop. The best education happens in this way. The golden rule in learning is: actions – however slow – speak louder than words.

Getting him started

Let's suppose you have made out cards, playwiths or charts for these games. You naturally want your child to play. Rather than attempt to persuade him to play, leave the cards lying about on a table. When he picks them up and asks what they are, reply casually: 'Oh, just a game.' His curiosity is then likely to be aroused. Never let your enthusiasm bubble over and become a barrier between the child and the game. As every reader knows, there's probably nothing more dampening to one's interest than someone else's over-anxious zeal.

A father may make up his mind to play math with his son, but he doesn't know how to get going. His girl of four is playing trains with blocks. 'Let's do something very interesting – all about math,' he says. He doesn't casually suggest that she smarten up her train (or barge or rocket or whatever) by putting all the greens together, all the blues together and so on, and so forth. No, he positively dragoons her into 'sorting'. So she immediately wants to play something else.

Or a mother is anxious to help her daughter count. Her little girl has arranged six ash leaves to look like the petals of a flower, with a chestnut on top of each leaf. 'Now we're going to count: how many leaves are there? And how many chestnuts?' she asks. 'Six,' her daughter answers correctly both times. The mother spreads the chestnuts out in a row, and asks: 'How many chestnuts are there?' 'There are more chestnuts now,' says the girl. 'No, there aren't,' her mother corrects her. 'How *could* there be more?' The little girl can't explain: to her there just *are* more. (Most children below five feel in their bones that numbers depend on how things are arranged. It's the way they think: they see things that way.) If the girl sticks to her guns, she gets her first taste of being 'told' things by grown-ups. She slips away to do something she *can* understand and enjoy.

Discipline

Free choice, like curiosity, is a mainspring of a child's interest. Though really you're only as free as you *think* you are! You remember the story of Tom Sawyer having to whitewash a fence. He doesn't want to do it. Then he hits on the plan of 'allowing' other children to help him if they pay him a dead mouse or some other 'valuable'. The helpers are glad to pay for the pleasure of whitewashing a fence. This is not to say you need go as far as Tom Sawyer in grabbing your child's interest. However, sometimes a little stage-management can help things along smoothly.

The pendulum has now swung in favour of giving a child a free hand in the home. A couple of generations ago children were ruled with a rod of iron. Today we want to give them their heads. But that doesn't mean they should be in charge. For one thing, they would hate it. Children like to be firmly led. When they hit back, half the time they are only seeing how far they can go. Mostly they really want to be good and do interesting, sensible things. For another, children become quickly bored if left to do as they wish. The problem is, how do parents lead a child to think for himself and learn? In short, how does one *educate* a child?

Books and books have been written about discipline. So we won't add any more to the simple rule of being firm, meaning what you say, and sticking to your guns. Children respect parents for taking a firm line. But there are subtle, 'undercover' ways of getting a child to 'behave' – that is, behave in a way the parent wants.

A father may play with his child for ten minutes each day, with his mind on something else. If the child doesn't get his full attention, he'll make sure he does – even if it means being naughty. It seems certain that children would prefer a good spanking to being totally ignored. There may be the added consolation that in spanking his child the parent has gone to some pains to do something for him. But generally it's better for the parent to leave the role of authority to the teacher. The father, especially, can be the 'good friend in need'.

Don't try to trap him

Parents can make too much of a 'thing' of playing games. Let's suppose a father who has read up the sorting puzzles (Set 2) is determined to make things go by the book. 'Put your dollies on the far side of the table and your farm animals on this,' he encourages his daughter. She wants to put big toys on the far side and little ones on this. If he insists on his way, he may kill her interest. Far better for him to go along with her whim. His interest will put her in a cooperative mood.

This is good psychology at any level. You know how important it is for a young executive to keep the director happy. Perhaps he has a new idea that he, the executive, is sure is bound to help the firm. Does he tell the director outright to change his ideas? Not if he has any sense. He suggests his idea in a polite, casual sort of way so that the director convinces himself it's *his*, the director's idea – then it gets taken up. It may be good practice for a grown-up to get his ideas taken up by his child – and to see the idea changed almost out of recognition by the child. (This happens with directors too!)

A mother who has read about mazes wants to play a maze game with her boy. 'Now make sure you find out who can get through to whom,' she demands. Her boy runs his pencil all over the page, crossing the maze walls. It's tempting for her to say, 'No, no. Not that way. Here, like this.' Her boy would then look round for something else to do. The most helpful thing to do is always to respect a child's 'No'. And then try another tack. In fact, it's pointless trying to force a child to think. He'll find a way of escaping unless he feels the choice is his.

Or a father who is keen to help his daughter's number work returns home from work determined to get going on the 'Mr Aladdin' Number game (Set 9). His over-enthusiasm sets up a barrier in his little girl's mind. Her gaze wanders outside the window, and with it her interest. Suddenly she finds she wants a drink of water or to go to the lavatory or that she has a tummy pain. Anything rather than be trapped!

Grab his interest

Every salesman knows the value of a good line of patter, so why shouldn't the parent or teacher have one, too? Be friendly in a matter-of-fact way. 'Yes, it *is* a pretty color, isn't it. Is it your favourite?' (Incidentally, most people reply to a salesman's pitch because they're tickled that someone cares what they think!)

When your child objects to playing a game you can accept his 'No' or try an alternative 'line'. You want to play with colored shapes, but your child is being 'difficult'. 'I don't like blue rabbits,' he whines. 'O.K., maybe they aren't so nice. What do you *really* like?' If the little boy tells you what he prefers, rethink the game in terms of his choice, instead of 'blue rabbits'. You have to be on your toes which makes it all the more interesting for *you.* Like a good actor you have to be ready to play to your audience on the spur of the moment. Of course, your child may refuse to tell you what his favourite is; then you had better give up. Either try another game or let him go off on his own.

Again, the child may behave in what to you seems a downright ridiculous way. You have collected some sticks and stones for the ping-pong puzzles (Set 2), for your daughter. She points to a white stick and says it's black. It's better not to give vent to your initial reaction of exasperation. A subtler technique may achieve more. (A grown-up can get his own way sometimes, too!) Poker faced, you ask, 'Well, why's that? Why do you think it's black?' When she tells you, you can make her feel she has your approval – all she probably wanted – by agreeing. That's the name of the game called education.

The end of a good game should be like the end of an episode in a well-made TV serial. Experience shows that the 'cliffhanger' technique works with a child's learning. When a game has gone down well with a child the parent is tempted to play another, harder game. It's far better never to outstay your welcome; just hint at the next game, then break off with a casual excuse: 'I've got work to do,' or 'I must get tea ready.' There's no need for the parent to feel the good work has been lost. Almost certainly the next day the child will ask about 'that game we just began'. At an adult level, you've only to think how a casual remark dropped into the end of a conversation was the one that stuck in your mind.

Untidiness in the house

Parents like their homes to look attractive, and untidiness understandably offends their eyes. It's only natural for a mother to want to keep the house spick-and-span, but how can she with a child about? (It's more suitable to talk of the mother now as she usually has the job of clearing up the mess.)

One way is to earmark a space specially for making a mess in – the child's own room, perhaps, or a corner of a living room. Here the child should be allowed to make as much mess as he wants with paper, glue, paints or crayons. This way the child's irrepressible untidiness is confined and need not spread all over the house. When the child is through with messing about, he must be made to tidy up. (Few young boys like tidying up, but most young girls are willing to help their mothers.)

If she says to her son of four, 'Tidy up that mess immediately,' it may put his back up. But if she helps him with it (she'll have to do it herself in the end) he may quite enjoy it as a game. And he will grow to like doing it because he feels it's a 'grown-up' thing to do. She can increase the fun by playing on the sorting ideas in Set 1.

When he goes to school

Some parents happily coach their children, help them to match and count things, to read figures, and know the names of colors. While others would run a mile rather than teach their children math. There's nothing to be gained by their trying to force themselves to do it. Much better to leave it to the schools to do a good job. On the other hand, there's no reason why they shouldn't play some of the games in this book simply *as games*. They can comfort themselves that there *is* math in the games; if it's not obvious, then all the better. What should the parent do about the child who comes home complaining 'Teacher's awfully strict'?

To say that teacher knows best and not to interfere is, on the whole, probably the best course. But it can be helpful to let your child know that teacher may not always be right. (Adults have been known to make mistakes!) Teachers by the very nature of their profession have to put on an authoritative 'walking encyclopedia' air. But many realize the value of admitting their mistakes and, more to the point, turning to advantage a child's mistakes.

Parents usually want to see pages of ticks on the 'sums' books their children bring home. Though this isn't always evidence of good learning. Now we don't want to give you the idea that a child shouldn't get his sums right. But if he never makes a mistake, he's never going to learn much. So there's no need to feel downcast about your child's crosses. Without letting him know, look over his work. See if you can tell where he's gone wrong. Then perhaps you can find a game here that may bring the error home to him. Take the case of the little girl who doesn't see that two blocks and three blocks amount to the same as three blocks and two blocks. It's no use exclaiming, 'Well, it just *does*!' Deep down, she'll give up, and opt out of math for good. It's far more effective to play it out with the blocks (see Set 9).

Keep in mind that it's wiser not to contradict what the teacher is doing, even if you don't agree with her. You are unlikely to change *her* view, and you may confuse your child. Of course, you can discuss such a difference of opinion with your child and let him know that you are always there. The teacher, after all, won't be.

Coping with the new math

The "new math" children bring home doesn't look a bit like the "traditional" sums parents remember doing at school. So they may not have the slightest notion of how to help their children. It's really a lot simpler than it looks at first sight. But it's hard to see the problems clearly without looking at its history.

The "new math" began in earnest in 1957 when the first Russian sputnik rose into space. Interest in math teaching flourished outside the Iron Curtain countries simply because Russia's mastery of space was thought to be based on superior schooling in math. Within years new math programmes had been created to face this competition. Typical were America's Madison Project, Canada's Sherbrooke Project, and Britain's Nuffield Project. Several wise and distinguished mathematicians devised new school programmes—Dienes (one of us) in England and Australia, Thwaites in England, and Davis in America, to name but three.

These programmes brought in their wake changes in teaching approach and lesson content which have benefited children and teachers alike. But it's not possible for math teaching to go through an overnight change—literally a revolution—without raising problems: some parents still feel rather in the dark (which is why there's a need for such a book as this). It's natural and right to cling to what you know—that's how civilization keeps going. But if you see traditional math for what it was, the change is easier to take, and you can see the value for making it. Traditional math amounted to a highly practical way of doing sums, and, as such, it proved itself many times over. But today there's a definite need for people who understand what math is *about.* The computer can do much of traditional mechanical math. The aim of the new math schemes is to put a child on intimate terms with the ideas underlying math. The new math has more relevance to other school subjects than the traditional ever did. Your child's first steps in geometry may seem a far cry from your own memories of "Euclid." The emphasis nowadays is on doing things and making things. You may sense this fresh approach in many of our games.

"Will the new math muddle up the old?" is a question that vexes many parents and teachers. When your child learns new math—even if not as part of a major project—he will meet games and ideas like those here. He'll learn about sets (collections of things) and the way things link up (see Set 6). Children still need number work, but nowadays they approach it in a friendlier, less cut-and-dried spirit (see Set 9). If they enjoy number work they are apt to do it better, know how to use it. It's not much good a young child knowing his tables and then asking, "Please Miss, is this a multiply or a divide?" This is a worrisome question for teachers. It's a sure sign that the child doesn't understand how to use his tables—unless he is bored and being difficult. The real point of learning is to be able to *use* what you know.

The new math has been devised to prevent this sort of thing—in fact, to explain traditional math. If it *does* muddle up the traditional math, it probably means the old ideas were based on rote-learnt mumbo-jumbo and not on an understanding of the process. At the very least, the new math shows up the flaws in the child's learning.

PART 3

The Games

SET 1 **Sorting Games**

Before you play these games with your child, ask yourself these two questions:

1. Is your child at home with the idea of what a thing is *not* – like an unbirthday for instance?

2. Can he sort things not only tidily and neatly but to some *logical* pattern?

If the answer is 'Yes' both times, then you need only skim through this set of games. If not, then you may find two novel ideas to entertain both you and your child. The next two games should answer Question 1; 'Trains', Question 2.

The 'What Are You Not?' Game

As a warming-up activity, ask your child what he is *not.* At first sight it may sound nonsensical to an adult but a young child regards the question with the same attention as any other addressed to him. With prompting – because he may not be used to such an open-ended situation – he might offer these possibilities:

'I'm not a tiger.' (Another day he may well be a tiger!)

'I'm not a train.'

'I'm not my sister.'

'I'm not you.' – meaning the questioner.

At a more sophisticated level, a child may recognize that he is not a part of himself:

'I'm not my nose.'

'I'm not my toe.'

And so on.

The point to stress is, the idea of what is not in a set is as important to a mathematician as what is in it. This game smooths the child's way to sorting things *consciously* into two sets as in the nursery or playroom sorting (Page 52). For example, a child might sort his toys into red toys and all the rest, that is, the not-red toys.

Tell Me What This Isn't

4+ to 5 years

A follow-up to the last game.

Hold up to view, say, a bath toy like a gaudy swan and ask what it isn't. One four-year-old, Paul, told us:

'It isn't a swan.'

We didn't correct him: casually we enquired the reason for his surprising reply.

'Because,' he said, 'it has a greenishy beak . . . and swans don't have greenishy beaks.'

We explained diplomatically that it was *meant* to be a swan. And the game continued.

Naming Shapes

4+ years

A four-year-old once challenged us to name three round things on our person. We named wrist-watch and button with speed. But then were stumped. He supplied the third thing: the eye!

You could set games going by asking your child to name three shapes on you (or in the room) that are square (accept oblongs), triangular, or wiggly, in turn.

Sorting by Use

Sort things into sets according to what the things are used for. Admittedly this is not a common way of sorting. Each player thinks of everything he can think of that is used for cooking and, by contrast, not-cooking –

cooking and, by contrast, not-cooking
eating
cleaning
at school
on picnics
on car trips

To liven up the game sort by likes and dislikes. To make any sense mathematically, the players should stick to their preferences – though tempted to chop and change. One might just as well switch rules in arithmetic when it suits you. It is good fun perhaps, but bad arithmetic.

In the Kitchen

If you can still get fresh fruit and vegetables, sort the good from the bad (the maggoty or dried up). Best not to explain; simply show and be generous with praise when it's done well. (Essential to set happy emotional attitude in your child towards learning early.) Sorting into equal lots in scale-pans or by eye is more advanced.

Sorting these things is helpful:

potatoes (throw out ones with too many 'eyes' in them)
runner beans (don't want stringy ones)
blackberries (pick out red, unripe ones)
flowers (sort out the dead buds).

Sort the cutlery drawer (with all sharp knives out of reach), into sets of knives, forks, spoons; then split those sets into big and little pieces of cutlery. That makes six sub-sets. This is quite enough for a four-year-old to cope with. Or sort the cutlery into 'best' and 'every day' . . . another, equally good way of sorting.

Or let your child decide on a way of sorting. Decision-making is a vitally important strategy today. You can join in if you like. But, if you do, try to get something wrong (put a spoon in with the knives, say). Your deliberate mistake will test if he knows what he is doing – if he does, he should correct you.

In the nursery or playroom. It is a constant problem to find ways of coaxing a child to tidy up after playing. The solution lies in turning the tidying-up chore into a game. First you need several boxes, shelves in cupboards or cubby-holes for storing toys and books rather than one 'hold-all' box.

Big Box, Little Box

Put all the soft toys in one box and all the hard toys in another. Instead of boxes you could use shelves in the closet and a cupboard and so on. For example,

the 'soft' set: teddy bears
paper hats
picture books (paperback)
rag dolls
cloth or blanket (sucked since babyhood)

the 'hard' set: toy cars, trucks, planes
wooden dolls
clockwork toys
marbles
wooden blocks and cubes

Problem: What does one do with crayons? Are they hard or soft? The child must make an off-the-cuff decision: this is good practice in decision-making.

Then continue sorting this way:

red toys and not-red toys

The second set contains toys of every hue *except* red. The words 'not-red' are a tag or label for that set.
Then try:

blue toys and not-blue toys
green toys and not-green toys
soft toys and not-soft (that is, hard) toys – as above.
long toys and short toys . . .

Guess What

The child sorts toys into two boxes or places but he must not say how he's done it. The adult then has to discover the sorting rule.

Squaring up

A quick-as-a-flash method of tidying up. Simply square up books, jigsaws, paints, comics to the table's edge, square up boxes with the walls of the room. It's surprising how much tidier the room looks.

The math behind it. Draw a large square on a paper sheet. Better still, cut one out of cardboard (see the picture). Ask your child what he sees. By the time he

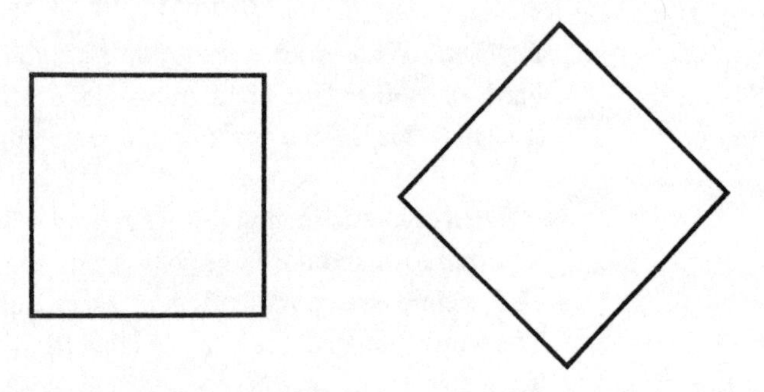

has learnt the word 'square' (at five) he'll call it that. Now turn the paper (or card) round until the shape looks like a 'diamond'. He will no doubt call it this if he knows the name. After much experience with shapes, he will see that the square is still a square whatever position it's in. But that comes later. The same applies to a triangle shape.

One of us recalls holding up a large equal-sided cardboard triangle and asking a girl as old as ten the name of the shape. 'Triangle,' she replied when it stood with its base on the table. But when the triangle was tipped partially round, she was at a loss for a name to give it. We asked her what she was called sitting down, then standing up and also running. No, she agreed she didn't change her name as she moved around. Then the penny dropped: she recognized that a shape doesn't change when *it* is moved about either.

Bottle in Glove

4+ years

This is not in any way a sorting game but we put it in as much to surprise as instruct. You may be amazed at the result yourself.

Things needed. Paper bag or duster or tea-cloth
Screw-top jar
Water.

Half fill the jar with water. Screw on the lid tightly. Drape the bag or duster over the jar. Cant the jar slightly to one side. Ask the child to draw the jar (or give him an outline of it) and the water level inside. Tip the jar this way and that; even lie it flat on its side. Ask him to draw the level each time.

Most young children, up to six- or seven-years-old, will draw the water level as shown. Then pull off the

my example
his copy
glove on
glove off

'glove' and let him see the level. Get him to see how his drawings compare. The sort of drawings young children make are shown in the picture.

Fishing

4+ years

A thinking game using the idea of a set. To be played with a group of children.

The idea. To fish for children according to some obvious quality. When you have filled your net, you have a set. The children have to guess what the quality is.

Play. Begin by announcing: 'I'm looking for some rather special fish.' Then 'fish' out, without explaining why, a few boys. 'What sort of fish am I going to catch next?' The children think what sort it will be. If they say 'another boy' they've found out. If not, go on till they do.

Then ask: 'Who are the children *not* in my net?' With luck, the children may declare: 'All the girls.'

Next you choose everyone with a party hat on; some hats are bound to have fallen off, so no fear of netting all the fish 'in the sea'. Or you might choose everyone wearing something red (or blue or green, and so on). Or everyone wearing a red party hat (or a blue one... and so forth). Or with blue eyes or brown eyes, red shoes, black shoes, or sandals or lace-up shoes. Avoid at first anything subtle – such as everyone who lives in flats (if you know). Once the game is going, hand over the 'net' to a child.

Detectives

4 to 5 years

Follow-up to last game.

The idea. To make a set of children who have *two* qualities. The children have to be detectives and find out what the qualities are.

Play. Choose all the girls with bows in their hair. We'll assume not all the girls are wearing bows. Once you have got your set of girls wearing bows, pose this puzzler:

'Look at my lot of people. You be detectives and tell me what is special about them all! The moment someone comes up with 'They've all got bows on,' tell them this is a *vital clue.* (If a child suggests your set is made up of all girls, you can shake your head, indicating the other girls without bows standing around.)

'You've got a clue,' you say. 'Each has a bow on. What *else* do you know about them?'

Your detectives should arrive at the same conclusion as common sense would have led you to – that they are all girls with bows in their hair.

Guess What I've Got

4 years

A child-like version of that old favourite 'Animal, vegetable, or mineral?'

Play. Suggest to your child that he finds a smallish household thing or toy and pops it in a bag or wraps it in paper or a duster. (We'll say it's a round hazel nut.) Now he lets you and any other children see the wrapped nut. You have to work out what it is by asking questions.

'I've got something round and brown and we haven't seen it for a long time,' he chants. 'Guess what I've got!' (This is an actual game played in an author's home.)

FATHER: Is it a biscuit?

CHILD: No.

FATHER: But you said it was brown and round.

CHILD: It still isn't a biscuit.

FATHER: Oh . . . Does Mummy like it?

CHILD: Yes. And it's white inside.

And so on, until someone guesses nut.

Suppose *you* play the 'Guess what I've got' role. With four-year-olds you could let them feel the thing inside the duster. With a very obvious shape – such as triangular, square, round, box-like or cube-like – you needn't tell them the shape.

Sticks 'n Stones Playwiths

5+ years

Go out and find real sticks and stones in these sizes and shapes:

9 sticks: 3 large, 3 medium, 3 small

9 stones: 3 large, 3 medium, 3 small

At each size, there is 1 black stick (or stone), 1 white stick (or stone) and 1 red stick (or stone).

The playwiths are shown in the picture overleaf.

If these playwiths are hard to come by, use buttons instead of stones and pencils instead of sticks. Coloring can be shown by sticking colored paper to the playwiths. Now you are ready to go.

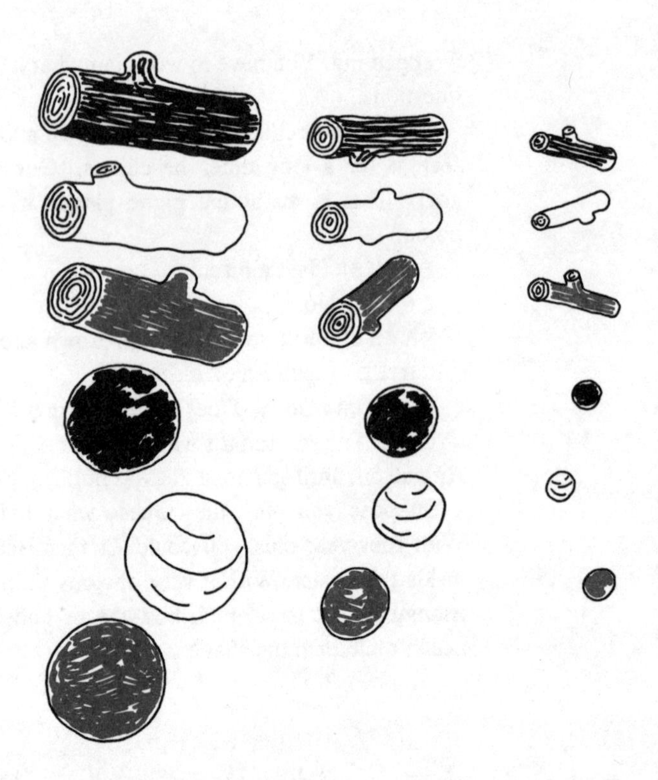

Dinosaur Nests

An ideal game for a picnic.

The playwiths are, we'll say, dinosaur eggs! Dinosaurs are a favourite with most children.

Put in a nest (or *set* as a mathematician would say)

a large black twig
a large black stone
a large red stone.

'What other eggs will go in this nest?' you ask. Anything large. Each child can place one other large playwith if he has spotted the rule.

You dole out a leaf as a reward for every correct playwith placed. A child can win two leaves if he can *say* what the rule is: '*Large* eggs.'

Other obvious 'nests' to build:

red eggs
small eggs
medium-sized eggs
white eggs
black eggs.

A player can earn two more chips (leaves) by making up another similar nest to match. Thus, you start off a nest of 'large' eggs. The other similar nests must be 'medium' and 'small'. Or you make a nest of 'red' eggs. Other nests that can be made are with 'white' and 'black' eggs.

Ping-Pong Sorting

This combines the ideas of all the earlier sorting games in one super-game.

Cast the playwiths down on an oblong rug. Sweep, say, the white sticks and stones up one end, the not-white (black and red) ones to the other end. Lay a ribbon or strip of card across the rug, to keep the playwiths apart. The picture shows this.

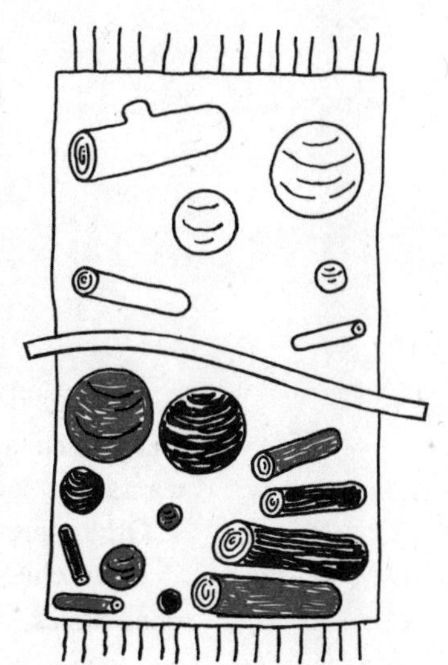

Now ask the child to sweep the stones to the left side of the rug, the sticks to the right . . . without letting the white and non-white get mixed again. (A common mistake children make.) The picture shows the ping-pong table or tennis-court arrangement.

You could ask a five- or six-year old to count how many in each corner of the tennis-court. But only if he wants to.

Other ping-pong tables are shown in the picture.

Your child can doubtless invent a few more ping-pong tables.

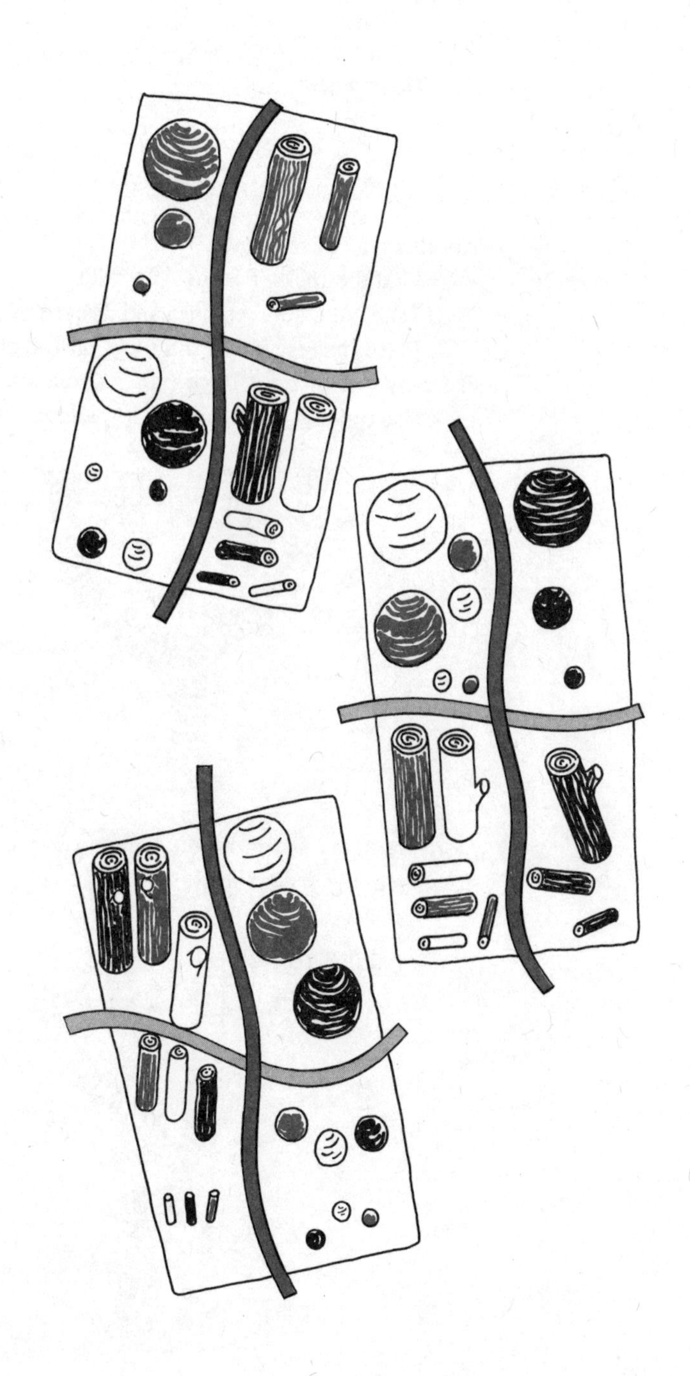

Trains

5–6 years

Another way of sorting.

1. Find the sticks 'n stones playwiths.

2. Take out all the red ones and discard them.

3. Place the rest in the shunting yard *A* on a sketch like ours drawn on a large paper sheet. Pretend they are trains running on lines as in the picture.

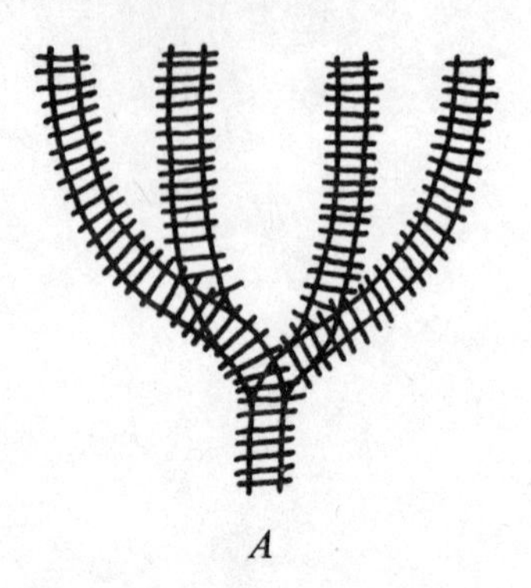

A

4. Mark the signposts on separate bits of card (or draw on to the paper sheet).

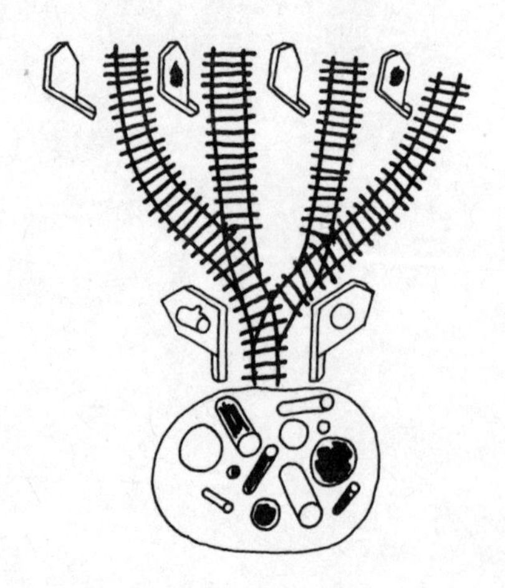

5. To suitable sound effects, perhaps, move each playwith up the first branch lines.

Follow the sign posts.

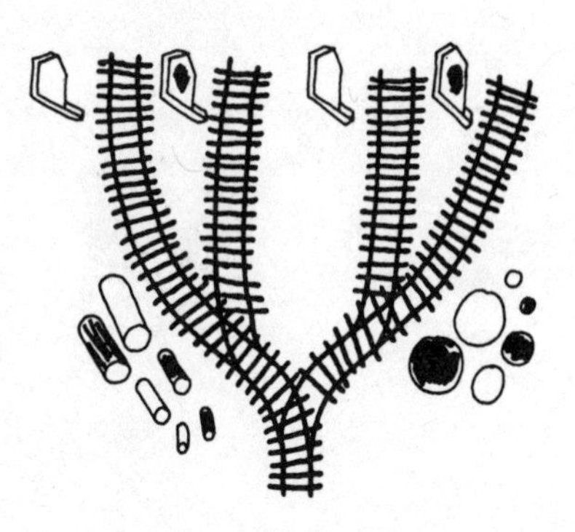

6. Push the playwiths along the next branch lines to the ends of the tracks.

Follow the signposts.

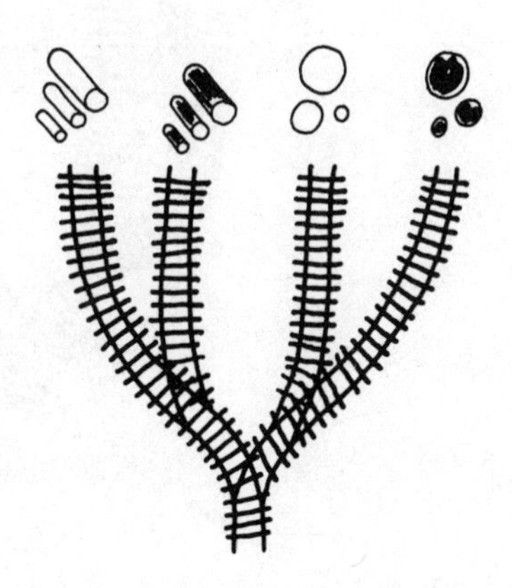

More trains

Ask your child to sort the playwiths along the same tracks with *these* rearranged signals:

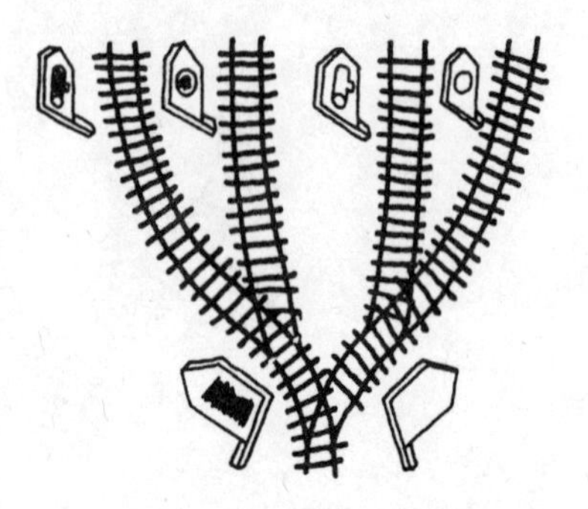

Make the train 'signals' look like this again:

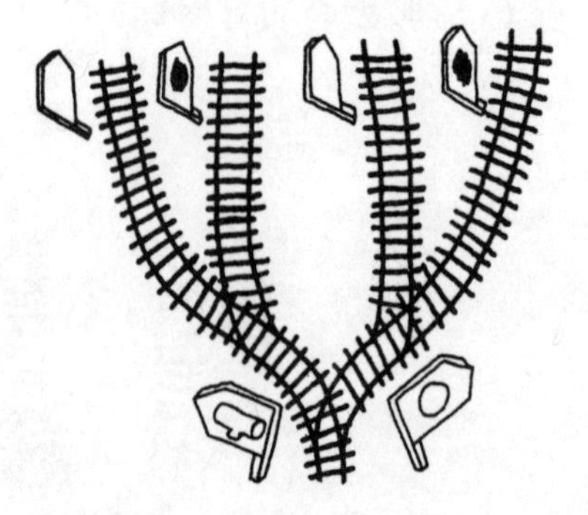

Now draw *this* plan on a large sheet of paper.
Send the playwiths past the first junction.

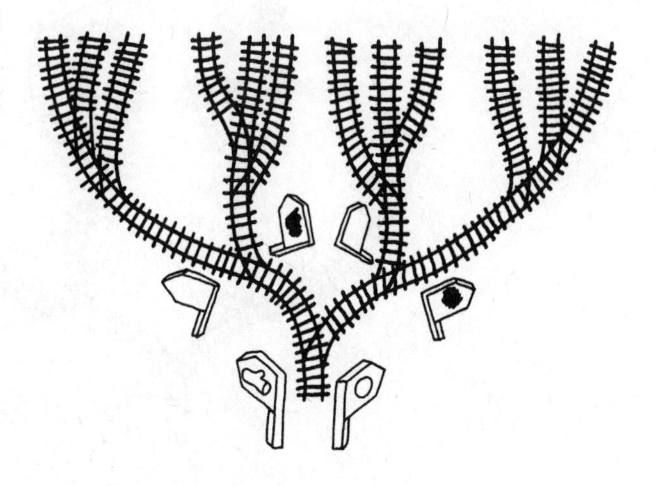

Send the playwiths past the next junctions.
You have this:

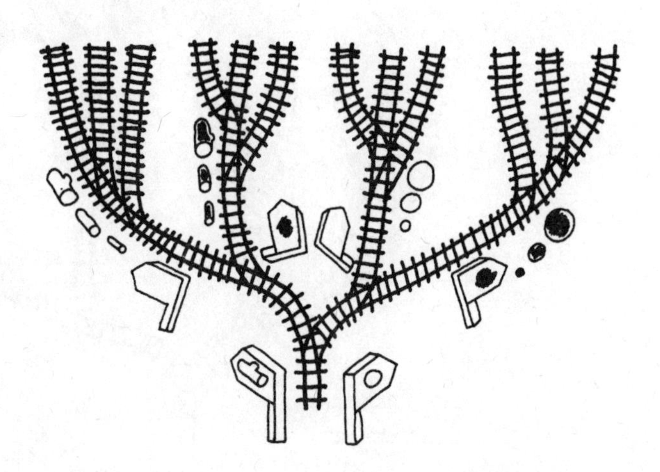

At the last junctions, move the big playwiths to the left, the medium straight up and the small to the right.

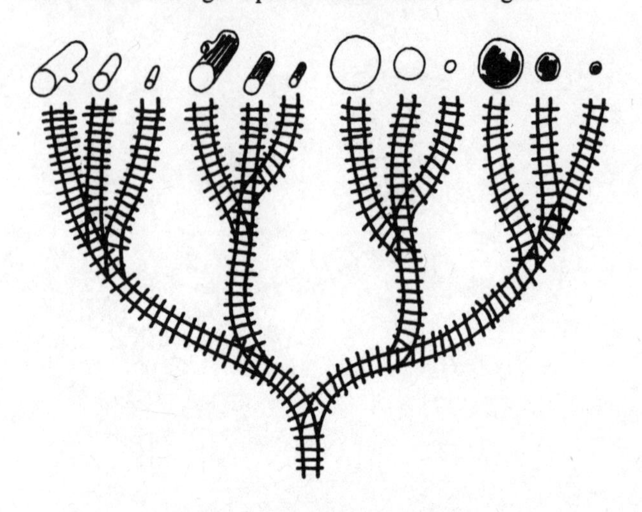

Now all playwiths are shunted to the end of the line. If you can, re-arrange the signals to get the playwiths to end up this way:

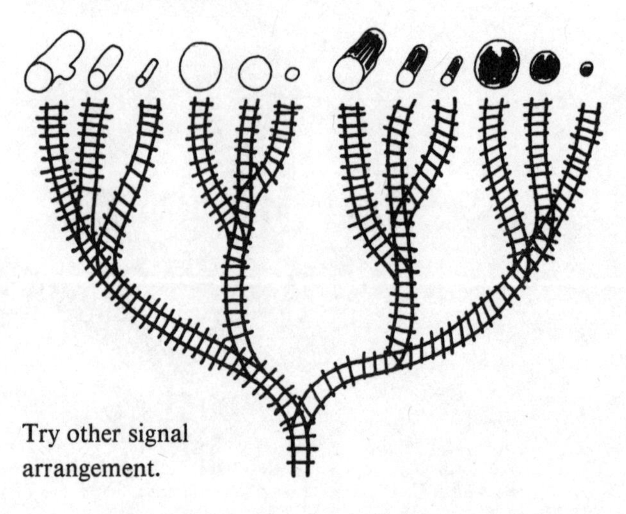

Try other signal arrangement.

SET 2 **Ping-Pong Puzzles**

Let's say your child has sorted his toys on a table-top two ways: for instance, into hard and soft toys *and* at the same time into dolls and animals, as in our ping-pong table picture:

So far so good. And with a little ingenuity we can turn this sorting arrangement into a genuine thinking problem – a Ping-Pong puzzle, as we call it. First, hide under the table all the soft animals, leaving a gap in the bottom right corner of the picture:

Now you can pose a ping-pong puzzle to your child:

YOU: Pick any animal on the table. (He points to one.) Then it must be . . . what?

CHILD (feeling): Hard.

This is correct. In logic we could write:

'If the toy is an animal, then it must be hard.'

Or, more succinctly,

'If animal, then hard.'

Here's another puzzler about the same picture with the very same gap in it.

YOU: Pick any soft toy. Then it must be . . . what?

CHILD: A doll.

In logic it would go:

'If soft, then doll.'

The interesting thing is, without the gap there would be no chance to reason. Try it and see! Replace the soft animals in the bottom right corner of the table and ask the first question again.

Question. Pick any animal. Then it must be . . . what?

Answer. (correct) Hard or soft: you cannot tell.

As you see, a perfectly correct answer, but one that tells you no more than you already knew. The gap may be likened to the clue in a detective story. Your reasoning must take the gap into account. If it doesn't, you still won't be able to reason. For instance:

Question. Pick any doll. Then it must be. . . . ?

Answer. Hard or soft – either: you cannot tell!

In other words, you are back to the situation where there wasn't a gap.

Here are several ping-pong puzzles to try with your child in the home. As a matter of interest, the first ping-pong picture was dreamed up by the author–mathematician Lewis Carroll who wrote the Alice books. As a result, in schools they are often known as Carroll diagrams.

Kitchen Puzzles

4 to 5+ years

Begin with the ping-pong picture of rounded or sharp, silvery or unsilvery kitchen utensils. Four gaps are possible – one at each corner. Here are the two ques-

tions you can ask about each gap with their correct answers. Make a gap in any corner – at the bottom right, to start with, if you like, hiding the saucepan and basin – that is, the unsilvery rounded things.

Question. Pick any rounded thing. Then it must be. . . . ?

Answer. Silvery.

It clearly wouldn't be much use saying 'Pick a silvery thing' or 'Pick a sharp thing.' Neither question would include the gap. The second question you can ask with this particular gap is:

Question. Pick anything that isn't silvery. Then it must be. . . . ?

Answer. Sharp.

Next leave a gap in, say, the top left-hand corner. (It doesn't always have to be the bottom right!)

Question. Pick any silvery thing. Then it must be. . .?

Answer. Rounded.

Question. Pick any sharp thing. Then it must be. . .?

Answer. Not-silvery.

Now take away the top right corner.

Question. Pick any silvery thing. Then it must be. . . ?

Answer. Sharp.

Question. Pick any rounded thing. Then it must be. . . ?

Answer. Not-silvery.

Finally, take away the bottom left corner.

Question. Pick any sharp thing. Then it must be. . . ?

Answer. Silvery.

Question. Pick any not-silvery thing. . . ?

Answer. Rounded.

That runs the gamut of all the puzzlers that can be posed round this situation. Of course, it is possible to make the same picture with the descriptions 'metal' and 'plastic' instead of 'silvery' and 'not-silvery'. And the ping-pong diagram can be distorted, made wiggly. The gaps in the pictures will look the same. The ping-pong picture might look like this to start with:

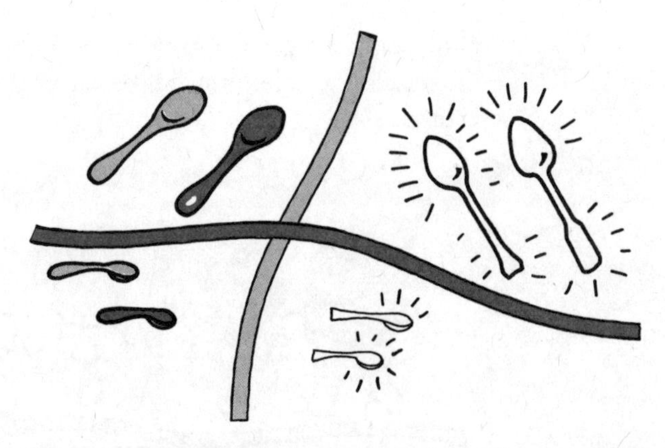

You can make up the same sort of questions, making a gap in each corner, in turn.

Nursery Puzzles

5 years

This makes use of the toys again. Make the ping-pong picture on a bed or a rug. Choose 'dolls', 'toys which aren't dolls' and soft and hard. (Instead of soft and hard you can divide the toys into red and not-red; or use any other color; and so on.)
Here is the ping-pong picture:

Now you have got the hang of the game, we present the puzzles in note form. Make a gap in each corner in turn.

Q: If doll?
A: Then soft.
Q: If hard?
A: Then toy.

Q: If soft?
A: Then toy.
Q: If doll?
A: Then hard.

That's all the gaps and all the puzzlers you can get from this picture.

Rainy Riddles

6 – 7 years

Glance a moment at this strip cartoon.

Read it *logically*.

1. Sam Jones wakes up one sunny morning. He looks out of the window and what does he see?
   'The road's wet,' Sam shouts.

2. 'Been raining,' he concludes.
3. Then Sam turns his gaze to the lawn.
   'The lawn's dry, though . . .'
4. Sam 'thunks', with think-bubbles above his head.
   'So it didn't rain in the night. . . .'
5. Farther down the road we see a sprinkler truck hosing the road!

The ping-pong picture of Sam's riddle looks like this:

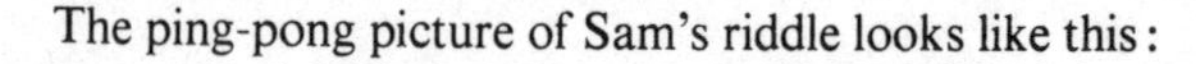

The gap in the top right corner shows you can't have a dry road on a rainy day.

Ask your child these questions and see if he replies with our answers.

Question. Pick a rainy day. Then the road must be...?

Answer. Wet.

and *Question.* Pick a dry road. Then it must be...?

Answer. Not rainy.

Wet 'n Dry Code

6 years

This is practice for what is to come in the next chapter.

Find the playwiths; discard all the red sticks and stones. You can read the playwiths as code signals for the Rainy Riddle pictures – like this:

a black playwith means 'rainy'
a white playwith means 'not-rainy'
a stick means 'wet road'
a stone means 'dry road'.

Now get your child to place black and white playwiths on the Rainy Riddle picture. Here's the answer:

That leaves the black stones without a home.
Next chapter shows you some code games.

SET 3 **Secret Signs and Codes**

Here are some novel card games: they are based on signs and symbols suitable for children a little older than those of the last set. Children (and adults) enjoy writing secret sign languages. When they can read, they like cracking simple codes. This is all to the good since they need to be familiar with symbols if they are to make headway with reading, writing and, later, math.

Tell My Secrets

6 years

Things needed. 12 blank cards (playing card size) or 12 old playing cards.

First, prepare the 'secrets' cards as follows. If you are using playing cards, first blank out their faces with adhesive labels. On them sketch the simple line drawings shown on the right.

Code chart. Make out this code card and place it prominently for all to see at any time during play.

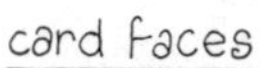

card backs

The game is for three or four players and is played just like 'Happy Families'.

Aim. To collect all the cards, by reading the code signs and asking for cards by their pictures. The winner is the one who gets all the cards.

Play. Deal out all the cards evenly among players with the cards lying picture-face down. Their backs with the code signs show.

Each player picks up his hand, holding it fanned out to show the backs of his cards (the code signs) to his opponents. He must be prepared to show the back of every card in his hand if asked. Player left of dealer opens.

Suppose Gary hasn't got 'a boy standing in the sun' (card 1).

But he sees Ann has a card with this back: he reads the code chart and asks her for:

'a boy standing in the sun.'

She must give it to him as he asked correctly.

Gary can now ask for another card of anyone, including Ann again. If he asks for a card and the player hasn't got it, it is that player's turn to ask. And so on, until one player has all the cards, and becomes the winner.

Pick Up Sticks

7+ years

A slightly harder version of 'Tell my secrets' for readers.

The game is the same but the code is different. Instead of shapes drawn on card, the sticks 'n stones playwiths are used. First discard the red playwiths. Here is the code chart:

big playwith = 'Bear'
medium playwith = 'Boy'
small playwith = 'Cat'
stick = 'stands'
stone = 'wears a hat'
black: 'by night'
white: 'by day'

Play. Lay out all the cards, picture side up, on a table. Put all the playwiths in a 'pool'. The adult can choose who plays first; play continues round the circle clockwise as in cards.

The first player takes out of the pool one playwith and puts it down in front of him. He works out what it means from the code chart and places it on the card he thinks it is code for.

If the adult (referee) agrees – or all the child players concur – then the player can take the card. He also keeps the playwith. In an even harder version, the playwiths are always returned to the pool.

Each player goes in turn and has one guess per turn. When all cards have been picked up, the winner is the player with most cards.

Cars and Houses

7+ years

This is the very first counting game in this book. It is a card game exactly like 'Tell my Secrets' but rather harder.

The front of each card has a picture of cars or houses; the backs have code signs which the players must read (that is, 'crack') in order to collect cards.

Things needed. Six cards (about playing-card size).

Prepare this set of 'Code' cards:

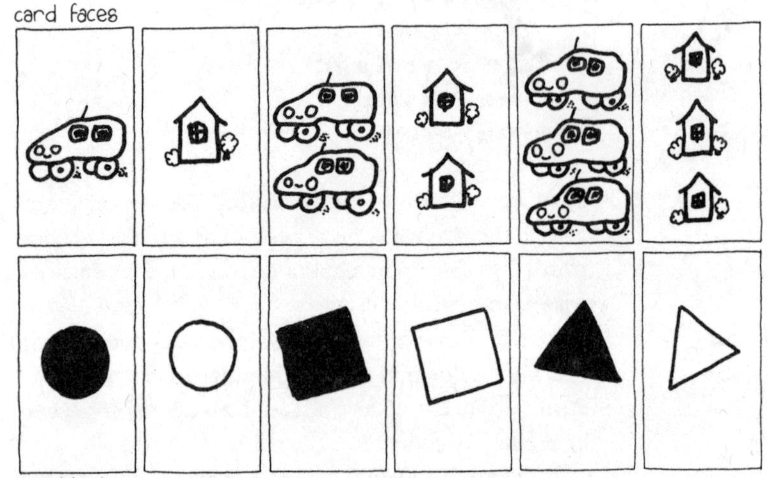

Code Chart.

Use this code chart (or draw a larger one). Place it prominently for all to see at all times during the game.

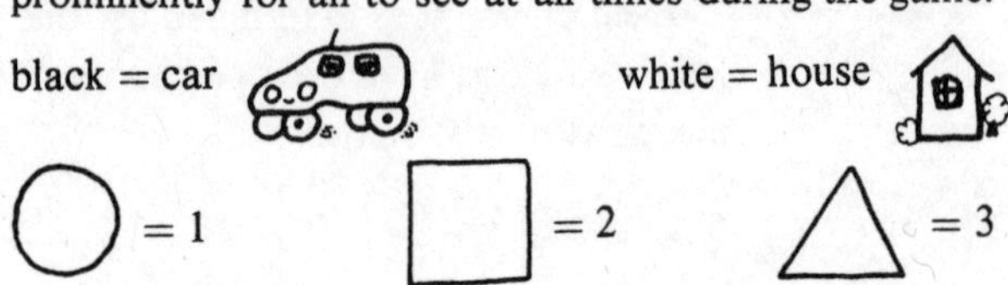

The game is for three or four players, just like 'Tell my Secrets'.

Aim. To collect all the cards by reading the code signs. Winner is the one who has all the cards.

Play. Players ask for 'one car', 'two cars' or 'three cars'; or 'one house', 'two houses', or 'three houses', by reading the code signs on the card backs where black signs indicate cars; white signs, houses; and so on.

SET 4

Drawing, Painting and Cutting-out Games

All his own work. Math is a kind of mental 'doing'. It's only when math is confined to scribbling meaningless signs on paper that it leaves no mark on the mind. One way of exercising the mind is through drawing, painting and cutting-out. Drawing and painting help in another way by capturing the fleeting moment. For children often need to work out what they feel through artistic expression; they may even get fears out of their system this way.

So how can you encourage your children's art? The good school tries to give a child a fuller, freer life, to help him enjoy himself. When a child enters infant school at three or four, one of the first things a good teacher lets him do is literally slosh about in sand and water. But there must be *some* rules, or the child will quickly grow bored.

Artists, too, need rules. Traditionally fancy-free, they are in fact craftsmen, and must follow the rules of their craft to create effectively. The well-known Swiss artist, Paul Klee, drew marvellous, fanciful pictures of a childlike quality. But his fancy was always kept in check by the rules of the medium.

Can one stimulate a child's artistic fancy and break the habit, formed all too soon, of drawing 'cardboard cut-out' type figures? We've all drawn cats made up of two circles, a large one for the body, a smaller one for the head, with a couple of triangles for ears, a long sausage for the tail and three straight lines each side of the head for whiskers. What we've drawn undoubtedly represents 'cat' to most people. But so does the word CAT . . . and much more efficiently! Such a drawing cannot be called art; it is picture-writing, no more. Through the centuries man has evolved from such picture-writing to the highly accomplished trick of speaking language and writing it with letters. Here are some more playwiths and games to kindle a child's imagination and break him of the 'easy-way-out' habit.

A child likes to feel that his pictures are all his own work. One is often tempted to improve a child's paper folding or to straighten a crooked drawing. It is better to resist the temptation. For the child must *feel for himself* the rules and limitations imposed by the playwiths (about to be described) before they can guide and help him explore his artistic talents fully. His control is not as good as an adult's; he must be given the opportunity to develop it for himself.

The Rainbow Toy

This is an artistic playwith that mathematically gives a pattern for coloring pictures of houses. Cut a square out of white card. Color both sides of each corner as in the picture. Back and front of each corner should now be colored the same.

Place the Rainbow Toy on a large paper sheet. Draw round it to form a frame. At each corner of the

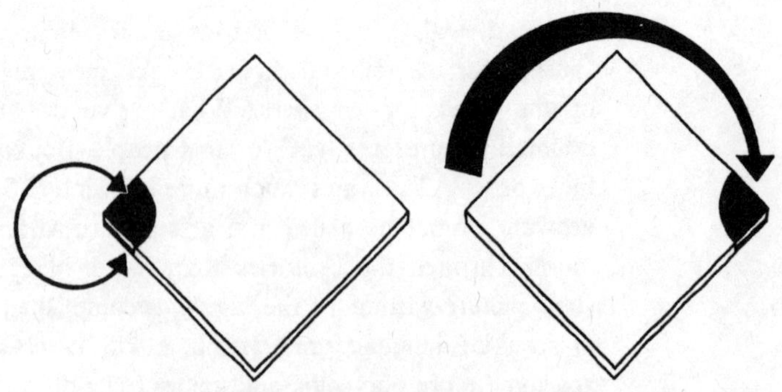

frame draw or, if he can read, write the words 'wall', 'roof', 'door', 'window'.

This is the Rainbow Toy. It selects the color-schemes for the houses the child is going to draw.

Have your child start with the Toy, as in our picture, which gives this color scheme: grey roof, red

wall, pink window and black door. This is the first house. See the picture at left.

Next he puts the Rainbow Toy down in its frame another way to give a differently colored house. By turning the Toy round further he can paint four different houses. By flipping it over, he can get another four houses. Few children believe that the second lot of four houses will be different from the first.

The Shape Toy

It gives a rule (like the Rainbow Toy) for changing the *shapes* of houses the child paints. Again, use the same large paper sheet with its symbols or words written at the corners of the frame. Cut a 'square' out of white card. At one corner on both sides draw a circle, at the next a square, at the third an oblong, and at the last a triangle:

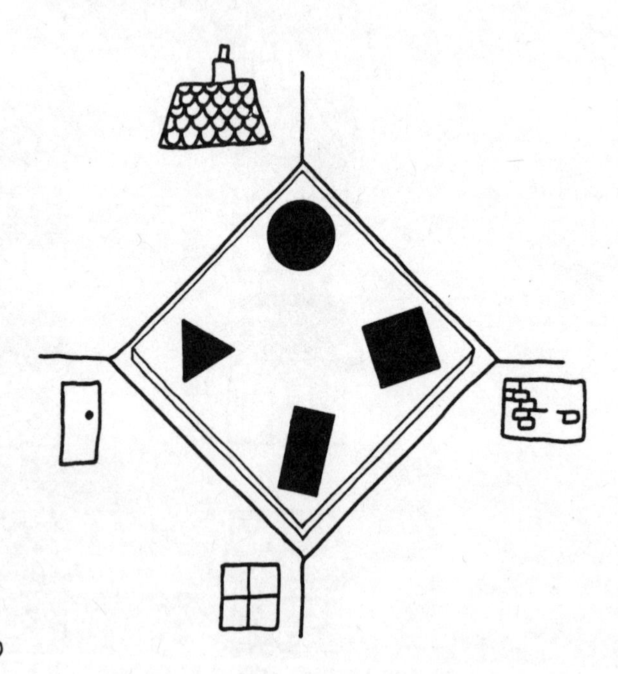

Children can achieve pictures of the wildest sort of house, like the one here got from the Toy in the position shown. Again, different houses result for the other side of the Toy. Coloring can be to taste.

There are actually 24 different ways you can combine four shapes to make up a house. But the Shape Toy obligingly reduces the total to eight, as did the Rainbow Toy, four for each side of the Toy.

The Three-Cornered Toy

This toy is a three-cornered piece of card. It gives a pattern for changing the color of the leaves, bloom and stalk of a flower. See our 'flower' got from the Toy in the position shown.

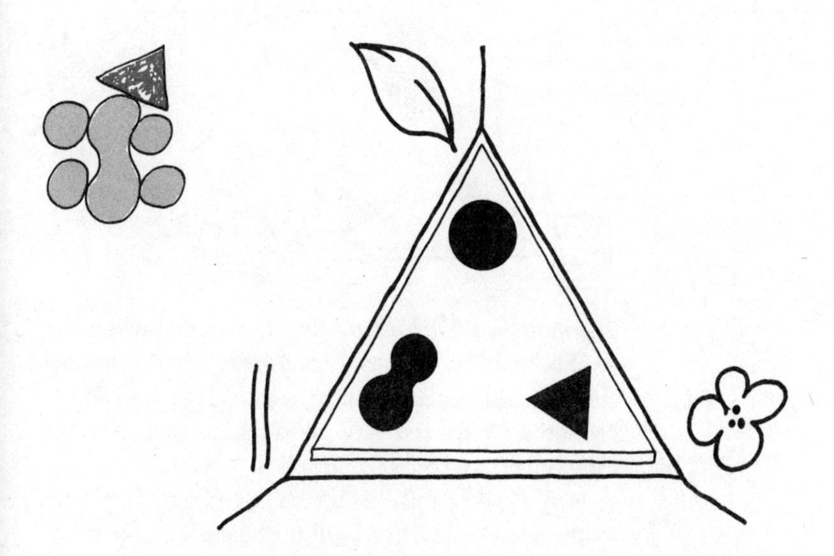

It is made and used in the same way as the other toys – except that a triangular frame must be drawn instead of a square one.

If you use both sides, you get six different flower shapes – some of them, admittedly, bizarre like the one

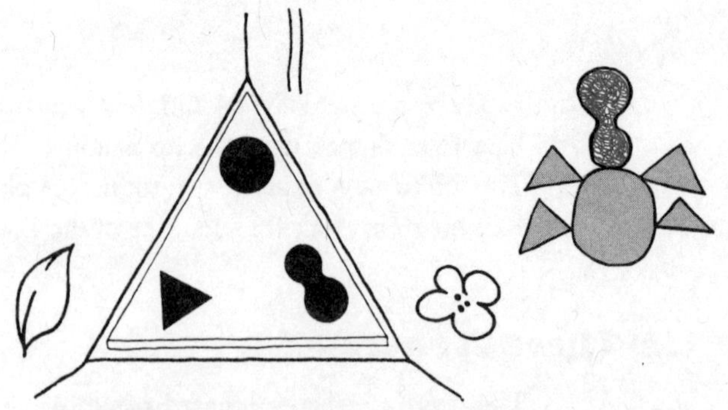

produced here by the Toy in the position shown, but that's the fun of it! Color it as you like.

The toy could also be used to change houses.

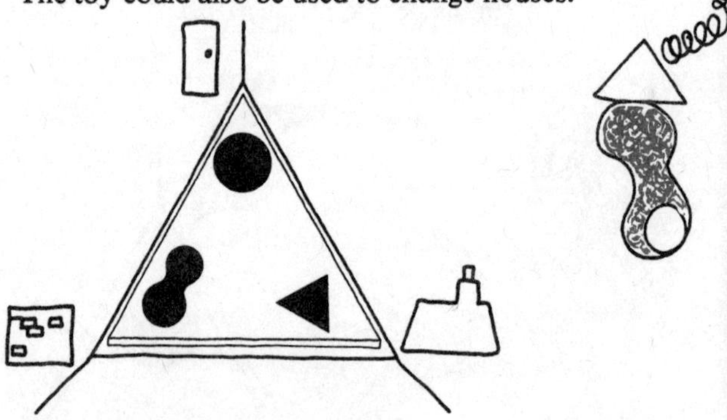

The math in the drawings. Skip this section if you like. A child may try to match his flowers with his houses. (In fact, you might casually suggest he try this.) But he cannot do it for the very good reason that there are eight houses and only six flowers.

Why is there this mis-matching? It stems from the math behind the Toys. On the Shape Toy, the color-schemes go in threes; on the Rainbow Toy, in fours. To match every flower with just six of the houses means that the color-schemes of the houses should change by threes twice over – with two house styles left over. But they don't! This is the reason you cannot match every flower with every house.

Paper-Folding and Cutting

Cut a sheet of paper so that it's square. The first time, it is better to show your child how to fold the sheet. Fold it along the diagonal, then again and a third time.

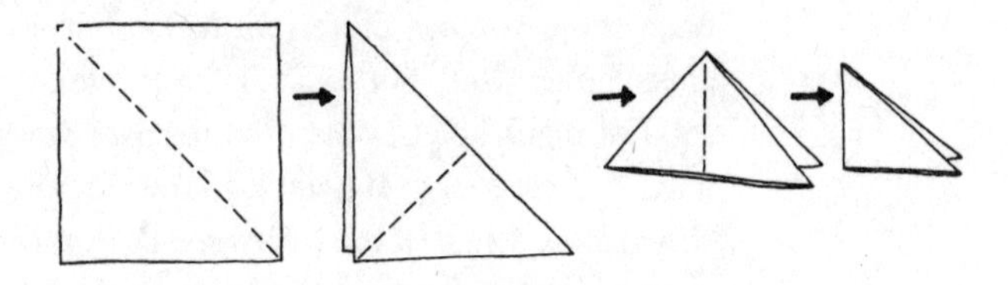

Now cut a hole through the folded paper from its edge. The hole can be in the shape of a circle, a star or diamond or even a dog. Ask your child how many holes he will see when he opens out the paper. Few children ever think there will be eight.

Next, hand your child the opened-out sheet with its lacy pattern of holes. Suggest he colors round each hole, then fills in each wedge-shaped slice round the hole, between the folds, with a different color. The spoke-like frame guides his artistic expression.

Mirror and Paper-Folding

Things needed. Thickish, unbreakable mirrors (plastic are best).

As a check to see if your child has cut the holes in the paper correctly, take a couple of thickish mirrors.

Place one mirror along a fold, one end at the centre of the paper square. Hold the other mirror along the next fold so that the mirrors form a 'V', their shiny sides facing in together. Look into the 'V' between the mirrors and you see all eight holes – though only one is real. Open the mirrors out in stages, like a fan. At each stage you see more real holes, but fewer looking-glass holes. But, you will discover, you always see a total of eight holes. When the mirrors are in a straight line, you see four real holes and four looking-glass ones – which is what you usually see in a mirror. You can have much fun showing this to your child.

Help him to see that 'left' and 'right' are always swapped in a mirror. Letters, as we know, are switched left to right in a mirror. Fold a sheet of paper down the middle. To the left of the fold get him to draw a sketch. Something lop-sided, like our picture, is best.

Hold a mirror along the fold. Get your child to look into the mirror; remove the mirror and ask him to draw the house and flower he saw in the mirror. He'll probably draw the house and flower he first drew. If so, say poker-faced, 'Take another look,' as you replace the mirror along the fold. 'Did you draw the house in the mirror?' It may take a minute or so before he sees his mistake. When he does, he'll be amused and draw something more like the looking-glass house. The daughter of one of the authors shouted with laughter when she made precisely this mistake.

Curiously though, she could draw and paint a butterfly quite well. She copied it from a real Red Admiral which had settled on a conveniently low marigold. She then painted the garden round her butterfly portrait and added a sketch of herself drawing the picture! She didn't draw herself again on this 'sketch' in her picture, however. If she had, she would have had to draw an endless series of ever-decreasing sketches within sketches.

Window Pictures

Ask your child to fold the paper once only. (Most children will automatically, without being told, fold it 'square' across the middle of the longest length, which is what's wanted. There is a temptation to pre-fold the paper for the child. Resist it.) Get him to draw something lop-sided on one side of the folded sheet. Then he turns it over and traces the drawing that shows through. Now how many pictures are there? Try the mirror test on them.

Take an oblong sheet of thin paper. Typing paper or

greaseproof paper are ideal. Get your child to fold the sheet twice, making four smaller oblongs. Suggest he does the following:

Draw in one quarter of the folded paper something lop-sided, perhaps one of these:

a cat with its tail curling to one side
a bent flower
a pixie with a hat on one side of its head

– anything so long as it's lop-sided. Hold the sheet against a window-pane in daylight. The cat (or any figure) shows through. Fold the paper along one of the folds. Trace the cat on the folded-over bit. Unfold and fold in the other way. Redraw the cat. Then flip the sheet over back-to-front. Repeat the folding and copying until four cats are drawn on one side and four on the other. Since the first cat was drawn lop-sidedly, there are four *different* cats.

Finally, hold a mirror along one of the folds: the looking-glass cat should be the same as the one copied on the sheet of paper on the other side of the fold.

SET 5 Maze and Dance Games

Muscle memory. A boy we know was rated 'backward' by his teachers, yet he was quite capable in a practical sense. It turned out he was afraid to use his hands. He couldn't, for instance, name the shape of a round, plastic plate because, we found, he didn't dare pick it up and hold it in his hands. To tell him to sit still and use his eyes would hardly have made matters any better. What he needed was encouragement to handle things so that he could *feel* shapes. Such children quickly get the idea that they are dim and so come to hate math and other subjects.

Psychologists have given a great deal of thought to problems like this. They have made a special study of the way children learn. Bit by bit they have learnt a great deal: that children learn by moving about (they cannot, we now know, learn well when being made to sit still; if they are told to, they only fidget); that child's play is in real earnest – for him, it is hard work; that the stronger the link we forge between our fingers and our brain, the better we become at doing, among other things, math; that our muscles remember for us as our brain does; that we make pictures in our minds (which is all that math is) from handling real things; that we best learn math, as most things, in a down-to-earth, commonsense way.

Once Sir Isaac Newton, the scientist who discovered

gravity after an apple (supposedly) fell on his head, had a problem in handling space. He had a cat for whom he bored a hole in his kitchen door to get in and out by. When the cat had three kittens, Newton bored three more little holes in the door! A child's first appreciation of space is not much different from Newton's in this story.

Here are some games to show your child the grass-root facts of space. Many may seem glaringly obvious to an adult. But they are not to a child. The secret is not to tell the child: always let him discover for himself.

The next two games are about boxing-in space with walls.

Hunt the Bug

4 to 6 years

Things needed. Lots of boxes of every shape and size – from old cigarette cartons or sweet boxes, to giant packing cases or suitcases. All must have lids that shut. Two boxes should be easily recognized: stick red tape on one, green on the other.

In the red box pop a bead, a plastic box top, or scrap of tin foil: this is Billy Bug. (Your girl may want to call it a butterfly.) Get your child to pack the boxes, each within another. Unlike Chinese boxes two, three or more little boxes can fit side by side within a bigger one. Close all their lids. Billy Bug is snug in the red one. Put the lot into a jumbo-sized box, or, failing that, into a large drawer or cupboard.

Tell your child the Bug is in the red box but wants to move to the green box. Ask your child to change boxes for him with the fewest possible openings and shuttings. The rule is: he must shut each box after opening it. You can ask a six-year-old to count how many openings and shuttings he had to make altogether.

Dead-pan, tell your child the Bug has changed his mind: he wants to get back into the red box! The same game ensues. The aim: to see if the same number of openings and shuttings are needed. To begin with, a child will make many wasted moves. It usually takes him a long time to see that going from the red box to the green takes the same number of moves as going the other way. (Please don't give away this *apparently* obvious fact!)

Hole in the Box

4 to 6 years

Ask your child what is the fewest number of holes he must bore in the boxes to get the Bug from the red into the green box.

A child often makes more than one hole in a box and doesn't see that one is enough. (But don't tell him!)

Link this game with the first in this Set: ask your child if he has to open and close boxes more times than he bores holes when the Bug goes from the red to the green box. Actually, he needs to make *more* holes than openings and shuttings. How many more will depend on how the boxes are packed.

Mazes

5+ years

After boxing-in space with walls, we next want to divide up the walls themselves – that is, chop up surfaces. Which is what mazes do.

Two astonishing aspects of a child's appreciation of space have emerged from playing maze games. First, a young child does not see that the relation between two points in a maze is a two-way one. For instance, suppose two people A and B are wandering inside the Hampton Court Maze. It seems patently clear that if A can find a way through to B then B should be able to reach A. The 'accessibility' relation is plainly a two-way one.

But a child doesn't see it this way immediately, as we shall discover.

Second, a child tracing a maze with a gap (or a gate) in it behaves like a chicken trying to reach corn on the other side of a fence with a gap further along. The chicken doesn't think of going away from the immediate goal – the corn. Nor does the child.

We bring these concepts home to the child not by telling him but by showing him – the first in the very next activity, the second in 'Shut the Gate'.

Things needed. Giant sheets of paper
Colored fibre pens
Buttons or counters

All our mazes are actually very simple. Mazes need not be abstract in design: they can have the outlines of real things – trees, houses, cars and so on. Trace them off on to giant sheets of paper. The simplest maze you can draw is a sausage shape: it splits the page neatly into two (just as the wall split space into two). The Church-and-trees maze is really only an elongated, kinked sausage maze.

The basic rule. The lines stand for walls and you can *never* climb over them. That is, you must never cross a line.

Get your child to trace the sausage-shaped maze with a pencil or his finger. There's just one wall: so he'll never get in or get out of the sausage-shaped area.

Glance at any of the other mazes and say how many parts the page is split into. At first sight it seems impossible to say. But there's a trick you can use to find out: climb on the wall. (Rather, put your finger on a line.) Walk along it. (Trace the line with your finger.) Go as far as you can. Now do you return to where you began? (Does your finger move back to the spot it started from?) If so, you know for certain that that wall at least splits the page into just two parts, and two parts only. Get your child to trace the maze with his finger.

In various parts of the maze draw some 'boys' and 'girls' or an elephant, bird, monkey, fish or panda as we did.

Ask your child if the panda in the picture can reach the elephant *without crossing a wall*, can the elephant get to the panda? It's plain to you, of course, but – and this is very surprising – not to your child! He will have to retrace the maze with his finger to find this out. Actually he'll probably need plenty of maze-playing to hit on this (to him) remarkable fact. Do not tell your child – let him find out: then he won't forget.

Shut the Gate

5+ years

Ask your child where to put a gate in the maze here, to make it possible for the panda to get to the bird.

Some gates may completely change the picture: people or animals that couldn't reach one another before will now be able to do so. The gate connects new parts of the maze.

The presence of the gate makes it possible for the panda to reach the bird. As a check, most children trace the panda's wanderings and, initially at least, bypass the gate; to venture through seems to the child to be taking him right out of his way: moving *away* from his goal, the bird. The point is, in a maze actual distances and directions do not matter; accessibility is the key. A child may have to trace several such mazes to understand the idea of accessibility (A can get to B).

THE DANCES

6 to 8 years

These 'dances' are actually problems in handling space – very early geometry in fact. Seen in action, they have an intensely humorous quality. Though in mathematical terms the dances accurately mirror the properties of simple geometrical shapes – the triangle, square and oblong. This much is clear from a glimpse of the very form of the dances. What isn't so evident, however, is this: each dance involves the very same pattern of moves as one of the Pebble games (Set 10)! Yet could anything be more apparently different: a sequence of slow gyrations and contortions by a knot of children and a field game not unlike rounders? To cap the improbability of it there is yet another game with the same pattern of moves – an armchair game played with pebbles. The final straw of improbability is that all these games reflect in their moves the properties of the triangle, square and oblong – the sort of thing one learnt laboriously as part of Euclid.

It is precisely this many-faced aspect of mathematics that gives it its power and fascination. The temptation exists to tell the children about these similarities. Resist it: it is not only preferable to let them find out for themselves; it is the only way. Experience shows that children can find things common to complex situations like these only after a certain age – and after some practice. If your child is not ready for it, he will merely repeat after you what you expect him to say. He won't grasp what you are driving at simply because you have put the accepted words in his mouth.

When he does grasp the idea he will tell you about it: 'Dad, that painting game,' he will say excitedly. 'It's just the same as that dance we did!' What could be more different – painting pictures and dancing? As it happens, the cutting-out activity has the same pattern, too. In the Rainbow Toy game, eight different houses can be drawn; in the Square Dance (to follow) there are eight different positions for the children; in the cutting-out activity there are eight different shapes. We return to this theme in the introduction to the Pebble Games (Set 10).

Oblong Dance

A 'dance' for four.

Find a large *strong* towel or rug. (A sheet of newspaper may not do as it tears too easily.) Each child holds a corner of the towel. To limber up, get each child to move along the long side – a move we call a 'long' dance – and then along the short side – the 'short' dance, letting go of the towel.

That done, let's say the children whom we'll

call Earl, Sam, Liz and Bet, start in the places shown in the picture.

Get Earl, say, to do a 'long' dance. Ask everyone else to move with him – without tearing the towel. As well as being good housekeeping this proviso makes sense mathematically: a torn oblong is no longer an oblong.

Next ask the children to swap opposite corners. That is, Earl changes places with Sam, and Liz with Bet. Don't tell them, but it is just not possible to do this so that the towel fits over its starting position after the move. With a *square* towel, however, it is possible. This experiment highlights an important difference between an oblong and a square.

Square Dance

A 'dance' for four.

Begin with four children facing in, holding hands to make a square:

Ask Sam, say, and Earl to change places. Liz and Bet must stay put – although they may swivel round like spinning tops on the same spot. But –

Nobody may let go of his neighbour's hand,

And at the dance's end, no arms may be crossed over.

Each 'dancer' has to finish facing outwards. He will find his arms outstretched behind him, as in a swallow-dive, still holding his neighbour's hands:

Then ask the 'dancers' to return to their starting positions, with everyone facing in again.

Mathematically, this 'dance', and the following one, has the same underlying pattern to it as the Pebble Games. There is a temptation to tell children this extraordinary fact. It is better not to. For if children see the similarity of patterns on their own, they will have acquired a valuable insight into math for life.

Triangle Dance

A 'dance' for three, identical in principle to the 'Square Dance'. Begin with three children – Liz, Sam and Bet, this time – holding hands to make a triangle:

Ask Liz, say, to change places with Sam. Bet must stay put although she may swivel round on the same spot like a spinning top. But –

Nobody may let go of his neighbour's hand.

And at the dance's end no arms may be crossed over.

Each 'dancer' should finish facing outwards. He will

find his arms outstretched behind him, still holding his corner neighbours' hands.

Ask the 'dancers' now to return to their starting positions, with everyone facing in again.

Mid-dance, the 'dancers' will appear to be all arms and legs like the two children in the photograph.

Sand-Castles

7 years and upwards

A maze game, best played on the beach or in the snow.

Start by drawing in the sand a network of straightish roads, criss-crossing to form closed loops:

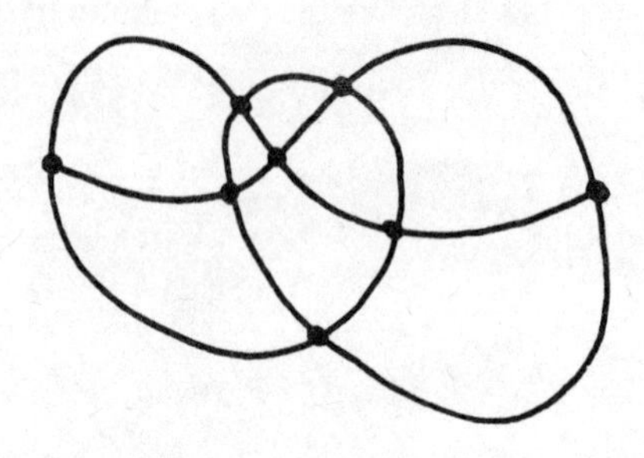

Rules. Each player, you suggest, is a castle-builder. The sand tracks are a network of roads. In the plots of land on each side of the roads they can build castles. Beyond the outermost road it is desert where they cannot build. The roads cut off a number of fields. Each builder, in turn, begins at any one of the cross-roads in the sand. He has a bucket with some shells (or pebbles) for money in it. At every cross-road is a toll-gate where he has to pay a toll of one shell which he lays in the sand near by. But he doesn't pay at the cross-roads where he begins or ends. When he walks along a track, he can build one sand-castle only in each plot on either side of the road (or he can draw a picture of a castle instead): for each sand-castle built he's paid one shell by the King (that's you!).

Aim. To make as much shell-money as possible. The way to do this is for the builder to plan his route carefully. For instance, he knows there's no point in going round three sides of the same field: he can only build the one castle on it.

Example of play. The picture shows two possible routes. Along the first route, the builder passes through

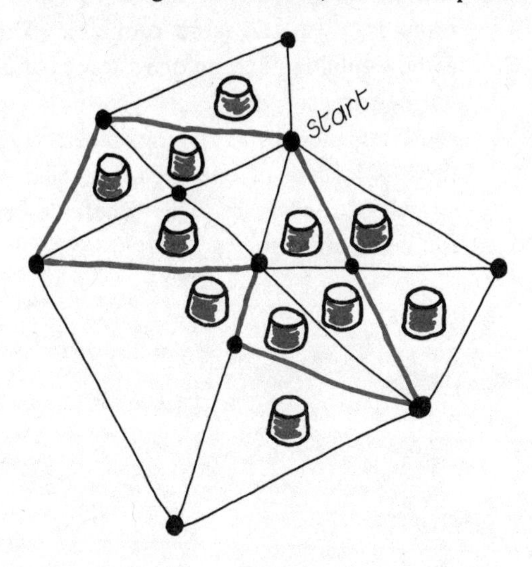

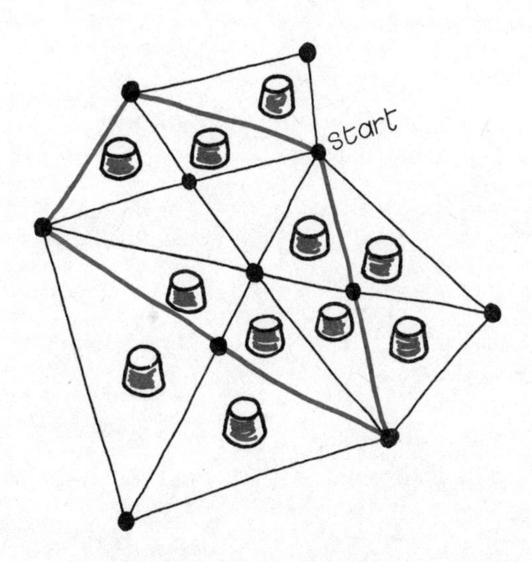

six toll gates. Of course, he does not pay where he begins or ends. So he pays out six shells, and he builds 11 houses, earning 11 shells. Profit: five shells. Along the second route he goes through only five tolls and builds 11 houses. Profit: six shells. The second route is, then, the more profitable. The secret is to avoid doubling back on one's tracks and to end where one begins.

A refinement for children of eight years or older: there is a 'fine' of one shell for going over the same stretch of road twice. This ensures even more careful route-planning!

SET 6 Linking-up Card Games

Linking up things and math. Math is about how things link up – how they relate one to another. The links may be numbers as in arithmetic or letters as in algebra or ratios as in geometry, or time-links like 'before' and 'after' (see the Time Games in Set 8). Understanding these relations and being able to communicate them are two important aspects of the new math, which is now part of most school's programmes – the Madison Project in the United States, the Sherbrooke Project in Canada, and the Nuffield Project in Great Britain, to name a leading few.

Probably the first relation a child encounters is the mother–child relation. Later he learns to recognize such simple relations as 'is taller than' and 'is shorter than'. Then he links these two relations together in such a way that he knows one is the reverse of the other: they are saying the same thing in fact. Later on he will use this facility for linking relations in math: he will see, for instance, that to say '7 is greater than 3' is the same as to say '3 is smaller than 7'. If he doesn't manage to, he will not get far in math. Consequently he may develop a so-called math block.

A crucial relation a child will need is the order relation. The next set of games is entirely devoted to this key concept. So we will say no more about it here.

A child's first conception of number arises from his

matching things – eggs with egg-cups, cushions with chairs, plates with table-mats, and so on. His first feeling for number stems from this activity: when he discovers that there are as many eggs as egg-cups. Surprisingly enough, this obvious-seeming fact does not appear obvious to a child until he has performed many such matching operations.

Another major step in his understanding of number is realizing how numbers relate or compare. Ask anyone how the number 5 relates to the number 6: they will tell you that 6 is 'one more than' 5. True enough. But the interesting thing is, a child will find such particular relations easier to grasp if he's met more general facts first. He might know by heart that $2+6=8$. But he might not know the general rule that adding any two whole numbers together always gives a bigger number. Give him the general first, and adding particular numbers becomes merely a special case of the general rule . . . and *for a child* that much easier to learn because of it.

These games give some practice in finding general rules and then in seeking particular relations from them: in moving from the general to the particular.

House-Building with a Difference

4 to 5 years To develop the idea of a relation of 'difference'.

Things needed. Colored building bricks (like 'Lego').

Ask your child to build a house all in red, say. Then suggest he builds another house, like it in every way – except that it is all white. Or better still, one child builds a red house, another builds a white one. 'How are the houses different?' you ask the children. After a few shots, most young children see what the difference is. It will take far longer to see it as difference of color in general, rather than a difference between two particular colors, red and white.

Later two children can build houses without looking at the other one. Urge them to try to make their houses look as different from the other as possible. When they have finished building, they can compare designs. They can count up differences between the houses. It may come as a surprise for a child to discover just how difficult it is to build a house completely different from his neighbour's! Instead of building, children can always draw pictures of different houses.

For special games on this theme for children to play, see the Painting games in Set 5.

THE HAPPY SETS CARDS

A pack of 24 cards. On them four things – bears, dinosaurs, dolls, cars: each in three colors – red, grey, black – and two sizes – big and small.

The cards provide children with experience of playing games about mathematical relations – of likeness and difference. They should ease the way to later work on number bonds, such as $2+3=5$.

Things needed. Large postcards
Colored felt pens.

Sketch the cards shown in the picture overleaf.

If you prefer, trace the pictures. The simpler and cleaner your sketches, the better – just so long as your children can understand them.

Like-It

5 years

A game for two players: a subtler form of 'Snap'.

Aim. To win the most cards.

Rules. Cards are played down face up together in turn. The player who spots the likeness in them picks up all the cards already played.

If he calls 'Like-it' incorrectly, the other player then picks up all the cards on the table. This rule is to deter 'wild' calling, without thinking.

Play. All the cards are dealt evenly between the two players who hold their cards face down. Each turns over one card from the top of his pack *together* at a given signal – a nod, say. The difficulty for young children is in keeping in time together. Cards can be alike in
shape: both cars, for example
color: both grey, say
size: both big, say.
They can be alike in one only or in all three of these ways. The picture overleaf shows some examples of play with their calls.

After the children have mastered this game, they should be ready for the next, rather harder version.

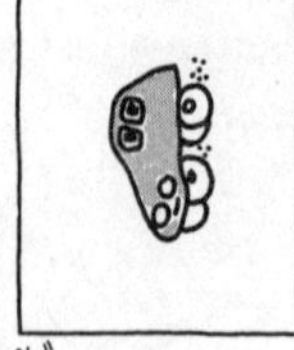

"like it"
(both grey cars)

"no call"
(alike in no way)

"like it"
(both black)

"like it"
(both big)

Quite-Like-It

5 to 7 years

A harder version of 'Like-it'.

Aim. To collect the most cards.

Rule. The player who spots two cards that are *alike* in only *one* way calls 'Quite-Like-It' and may pick up the cards on the table. To deter calling without thinking, the other player picks up the cards in the case of an incorrect call.

Play. The cards are dealt out evenly. Each player turns over a card together, as in 'Like-it'.

Cards must be alike in only one feature:

same size: both small, for example
same color: both black, say
same shape, kind: both dolls, say.

"quite like it"
(alike only in redness)

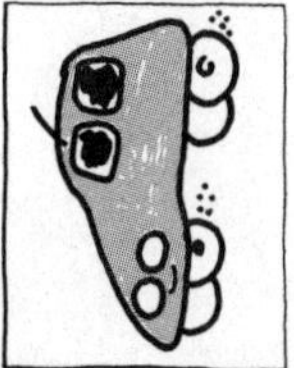

"no call"
(both big and grey)

"quite like it"
(alike only in smallness)

"no call"
(alike in no way)

Take the case of two dolls being played: they can't be both red, both grey and so on; nor can they both be the same size. Of course, if there's no similarity at all, neither player can call, as in the first game. The picture shows some examples of play with their calls.

Pelmanism

4+ years

A memory game for two to four players; it is played with the 'Happy Sets' cards like ordinary 'Pelmanism'.

Aim. To collect as many cards as possible.

Rule. A player may pick up two cards he has turned over if they are the *big* and *small* size of the *same colored* thing.

Play. All the cards are dealt out randomly *face down*

on the floor or a table. In turn each player turns over two cards. Suppose they are the big grey bear and the small grey bear: then he can pick them up. But suppose they were big and small but differently colored: he could not pick them up. Then the player must replace the cards face down where they were without moving any other cards. It is then the next player's turn. The skill is in remembering where the previously turned-over cards were.

Domino Cards

5 + years

A variant on Dominoes: for two. The cards are played in a line on the floor or table as for Dominoes. Next-door cards are not the same in some way – as in Dominoes – but are different in some respect.

Aim. To play all one's cards down on the table.

Rule. Next-door cards must have *one thing different* between them.

Play. All the cards are dealt out evenly between the players. Either player may begin. He lays down any card on the table. The other player then downs next to it, at either end, a card with a picture that has *one thing different* from the card(s) already downed.

For example, let's say the first card is a 'big grey dinosaur'. Next to it, can be played a 'small grey dinosaur', a 'big red dinosaur', a 'big black dinosaur', a 'big grey bear', a 'big grey doll' or a 'big grey car'. That is, six possible cards can be downed on either side of the first card; each has one thing different from the first card. The winner is the first player to down all his cards.

Any cards still in hand count against the player holding them. This rule only applies to those children who can and want to score.

Here is a line of play:

When a child has mastered this game, he should be ready to play the next, slightly harder versions.

Domino Cards (Two differences)

5+ years upwards

Rule. This game is the same as the last except that one can only down cards with *two differences.* These might be:

a change of color and shape
or a change of color and size
or a change of shape and size.

Here is a line of play:

Yet a third variation is to play for three differences, of course: a change of shape, color and size all at once.

Using the Secret Link-up Cards

Precisely the same games can be played with the cards for 'Tell My Secrets' (see Set 3). Instead of size, shape and color, you can play with color and the two shapes, the inner and the outer. Of course, as there are not so many cards, the games are not so interesting. It is interesting, however, that the Domino games can be played with the 'Secrets' cards face up or symbol up. Thus, if a child has difficulty in seeing differences between symbols, all he has to do is turn them over to read the same differences in terms of the pictures.

THE BLOBS CARDS

A pack of 16 cards with all the combinations of four different colored blobs.

With these cards several interesting games of a more mathematical kind can be played. They are specially designed to bring out the basic ideas of making sets and number bonds (and their underlying relations). These ideas your child will need at primary school.

The blobs are marked randomly on the cards so the child has to scan and 'read' the cards. This exercise is a useful adjunct to developing reading readiness. And it is a help for reading maths symbols; they are no more than simple signs on paper. It is what they *stand for* that makes them worth learning to handle.

Things needed. 16 postcards
Colored felt pens
or gummed circles (of any four colors).

Color the cards as in our picture. The blobs should be placed randomly. Or gum on colored circles randomly.

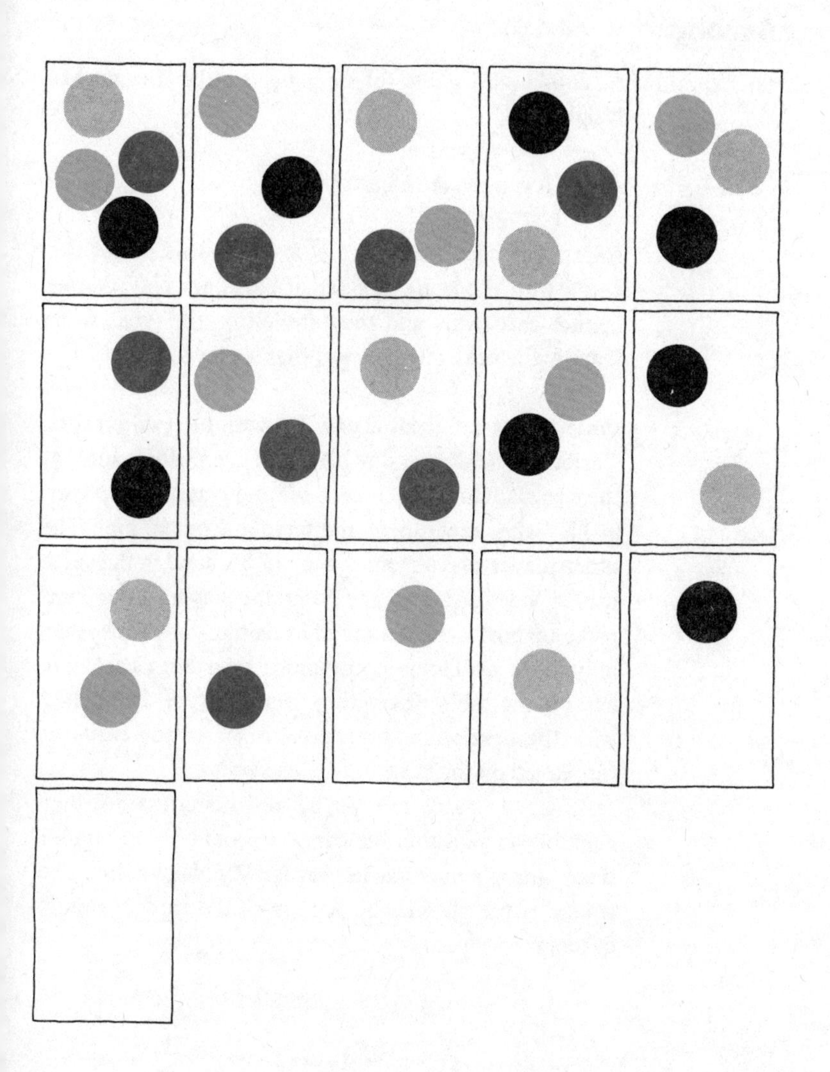

Here are several games a child can play with the cards, graded in order of difficulty.

Swoop

4+ years

A simple card game for two, played like 'Beggar My Neighbour'.

Aim. To win the most cards.

Rules. The card with more blobs on it wins. When two cards with the same number of blobs are played, then neither card wins and they are left on the table, to be won by the next winning card that is played.

Play. Cards are dealt evenly between the two players. Cards are held face down. The players don't look at their cards. One player leads off and plays the top card in his hand, face up on the table in front of him. The other player plays from the top of his hand in the same way. If one card beats the other, the player of that card picks up both cards. Sometimes neither card wins; they may have no blobs in common: then both cards are left on the table where they were played. The player with the next winning card picks up all the cards so far played on the table in one fell swoop.

If all the cards are played and nobody wins, then each player picks up the cards in front of him, shuffles them, and plays on, as in 'Beggar My Neighbour'. The winner is the player who scoops all the cards, or most of them.

Blobs

5 years

A card game for two players about sets and numbers again, like 'Beggar My Neighbour'.

This game helps a child build up the all-important concept of a *set.* Two aspects of the design of the cards give emphasis to this concept. First, the frame round the cards shows the blobs are completely contained within one set. Second, the child has to check that the winning card contains *all* the blobs on the other card and that no blobs are left out of account: both vital factors in making sets.

Aim. To win the most cards.

Rules. A card can take another if it has the same colored blobs on it *plus* some more blobs. Sometimes neither card wins. The card with four blobs, however, trumps every other card in the pack. The blank card, on the other hand, is taken by every other card. The reason: all cards have blank space on them as well as blobs, and the blank card is all space!

Play. As in 'Swoop'.

Two sample rounds show how the game is played. In the picture the card on the left wins the trick: not

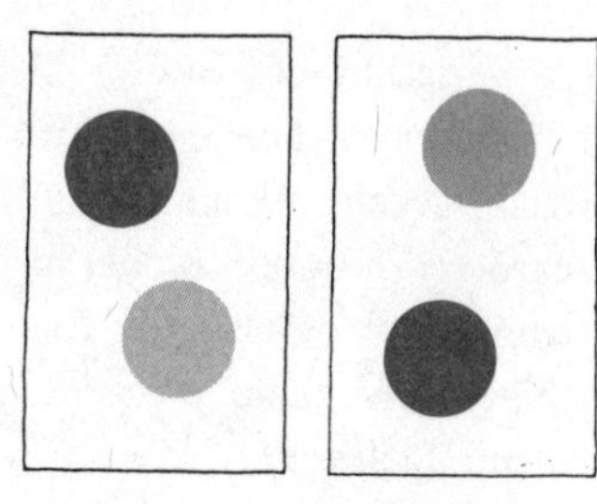

only does it have the other card's blobs, it has a grey blob, too. Any other color for the extra blob would have won, of course. And so would *two* extra blobs, only found on the four-blob card that trumps everything.

In the next picture the left card, though it has more blobs, does not beat the other: it doesn't include *all* the

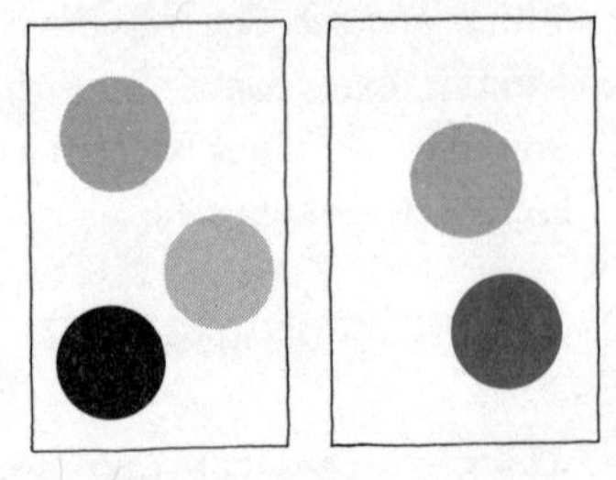

other card's blobs on it. Only the pink blob is common to both cards, which is not enough to win. So the cards are left on the table.

Tears

5 years

A variation of 'Blobs', like the *Misère* variation of *Twenty-one.* The card with the fewer blobs wins the trick. Again, all its blobs must appear on the losing card, with more blobs in addition. The blank card trumps all others; the four-blob card loses to everything. The game provides good practice for keeping mentally flexible.

Tricks

6+ years

An advanced version of 'Blobs' with this difference: each player may *select* from his hand which card to play next. Scanning one's hand of cards is that much more taxing for a youngster. The players win tricks as in 'Whist'. Of course, each trick will consist of two cards only; there will never be a pile collecting on the table to be won in one fell swoop as in 'Blobs'. The player who wins most tricks is the winner.

Domino Blobs

5+ years upwards

A variant on the Domino-sets game for two. It helps to build a child's readiness for 'take away' sums (differences).

Aim. To down all one's cards on the table.

Rule. The cards are played in a line on the floor or table as for Dominoes. Adjacent cards must have *one* blob of a different color. Examples below will clarify this rule.

Play. The cards are dealt out evenly between the players. The player holding the blank card may begin. He lays down any card on the table. The other player then downs next to it, at either end, a card with a picture that has one thing different.

For example, let's say the first card is a *pink and a black* blob. Next to it can be played a 'pink and a red blob', a 'pink and a purple', a 'black and red' or a 'black and a purple'. That is, four possible cards can be downed on either side next to the first card. Each has one color different from the first card.

Again, the number of blobs can be changed by one, while keeping to the same colors. The extra blob, whatever its color, makes the one difference. Or the one fewer blob could make the one difference – so a 'single black blob' or a 'single pink blob' could be played.

In short, the next door card can have *either* one more or one less blob, keeping the same colors of the remaining blobs *or* the same number of blobs, with one of a different color.

The winner is the first to down all his cards. (Any cards still in hand count against the player holding them. This rule only applies to children who can and want to score.)

Here is a line of play:

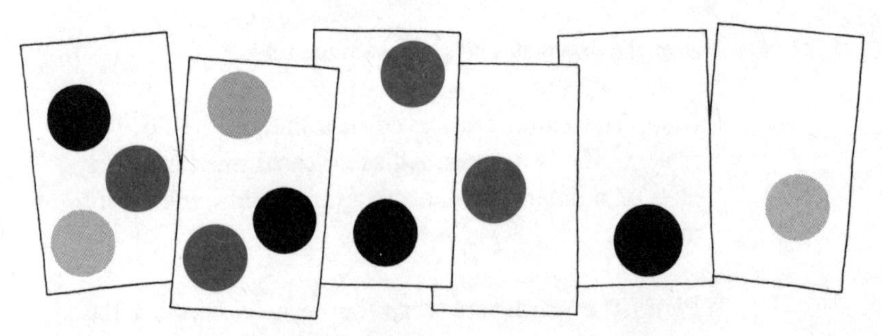

Domino Blobs (advanced version)

7 years

The game is the same as the last except that one can down cards with two differences.

These could be:

a change of two colors among the same number of blobs;

or a change of color *and* one more or one less blob;

or two more or two less blobs, the colors being the same.

SET 7 Ordering Games

Order to rule. 'Goldilocks and the Three Bears' is a story of orderliness in a child's world. The order in it is based on a rule of size – biggest first, smallest last: big bear, mother bear and baby bear each with matching porridge bowl, chair and bed. The apple-pie system recalls the well-ordered fantasies of the mathematician's mind. It also reminds one that the mathematician's world is quite close to children's rule-bound games. The first order-to-rule a child encounters is probably the A–B–C order of the alphabet and the 1–2–3 order of counting.

It is worth re-emphasizing that orders need not be rule-bound. The arbitrariness of order in math is rarely evident to the child even if he is able to count. In putting the Three Bears into an order, it would be natural to rank the Big Bear first; but equally reasonable to place the Baby Bear first – or Mother Bear!

There is a subtler side to order: if the winner of a horse race is disqualified on some count, there is still a first horse to take his place. Children find this aspect hard to conceive. They do not realize that order numbers (first, second, third, and so on) are different from counting numbers. This may be why the job of looking up words in a dictionary presents children with problems.

On the whole, a child has to cope with ordering

problems as best he can. Little formal training in ordering is given in schools. Yet with a little ingenuity it would be possible to turn everyday ordering situations into thought-provoking problems. Take, for instance, the rule about children filing out of a classroom. 'Smallest children leave first, tallest last' is such a rule. One could make the situation more interesting by adding a second 'ladies first' rule: when two children of the same height arrive together at the classroom door, the girl leaves first. Deciding on the order of their going requires considerable thought. We have not included such double-ordering games because they are too involved for infants to cope with.

King of the Castle

4+ years

A pencil-and-paper ordering game.

Draw a tree for your child or better get him to draw one. Draw it with four or five big branches drawn as uncomplicatedly as possible, one above the other.

Draw, each on its own branch,
a King with a crown, on the topmost branch
a Clown on the next branch down
then a Doll
a Bear
and a Cat on the lowest branch.

Ask your child such questions as the following:

'Who is at the top of the tree?' (Answer: King)
'Is there anyone higher than the King?' (No.)
'Who's on the lowest branch?' (Cat.)
'Is anyone lower than the Cat?' (No.)
'Who is higher than the Bear?' (Doll, Clown, King, all are.)
'Who's lower than the Doll?' (Bear and Cat.)
'Who's next higher than the lowest?' (Bear.)
'Who's next below the King?' (Clown.)
'How many people are higher than the Clown?' (One, the King.)
'Are there any lower than the Bear?' (Yes, the Cat.)

These questions proceed from the general (such as when we talk of higher or lower branches) to the particular (such as when we talk of the next higher or next lower branches). As we have mentioned, this fits in with a child's learning pattern.

The Clothes-Line Game

4+ years

An ordering game, with girls in mind.

Things needed. Ask your girl to draw, color and cut out paper dresses, like those printed in children's magazines and books for dressing up paper dolls. Two copies of each paper dress are needed.

Fix up two tiny clothes-lines in the form of strings hitched between two chairs. Suggest your girl hangs the clothes up to dry. She can use paper clips or bend over paper tags attached to the paper dresses.

The game. Put the clothes in some order – maybe from brightest to least bright, from favourite to least favourite and so on. The rule for the order does not matter, just so long as there is one. Now she should set up another, identical clothes-line so that the clothes on each line match up. Matching-up is vitally important to a child's understanding of number at this stage of learning.

She should reverse the order of the clothes on one clothes-line only. She may have to make several attempts before she understands what this entails.

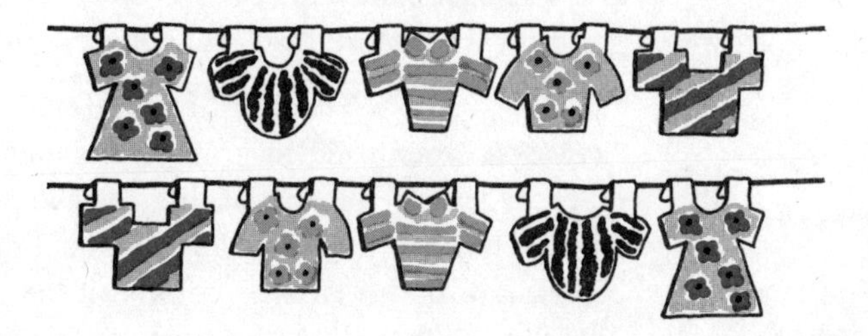

Then she should reverse the order of the clothes on the so far unaltered line. She may realize that both clothes exhibit the same order again.

Should a child take the game seriously enough to want to wash the paper clothes, enter into the spirit of the thing and draw a wash bowl with 'pretend' or 'make-up' water in which she can 'wash' them.

Suppose there are an odd number of clothes – say, 5 or 7 on each clothes-line. Then when the child reverses the order of one clothes-line, the middle dresses will continue to match up. If your child notices this, encourage her observation quietly. Suggest she make up a clothes-line without a middle dress. Finding a solution to this problem could represent a child's very first piece of mathematical research!

For your own information. There's always a 'middle-man' for an odd number of clothes – 3, 5, 7, 9 and so on; but never one for an even number – 2, 4, 6, 8 and so on. Try it and see for yourself. But don't tell a child.

Animal Grab

4 to 5 years

Things needed. Pictures of animals from newspapers and magazines. Stick the cuttings on light card to make playing cards. More than one copy of some pictures provides an opportunity to play Snap.

Play. Ask your child to arrange the animal cards in order of size of the animal. For instance, you might have pictures of a rabbit, dog, bear and tiger (in that order). A problem is likely to arise if the pictures are not all much the same size, as they are likely to be. The child may, rightly, want to order the pictures by their actual sizes. Thus a giant photo of a rabbit will be classed as bigger than a postage stamp portrait of an elephant. The solution is simply to agree on your ground rules before you start. Try these versions:

Any bigger animal 'takes' a smaller

The *next* bigger animal wins
A fiercer animal takes a more timid one.

Once an order of fierceness has been agreed, all players must keep to it.

Other cards and orders you can play are
People – put them in order of size
Aeroplanes – order by size
Cars – order of size or speed
Pets – order of popularity.

Stone, Scissors and Paper

4 to 5 years A modern variation on the old game.

Things needed. Grab bag
Colored counters or tiddlywinks
Ludo board.

The old game, you may recall, goes like this. Two players face each other. At a given signal – a nod or one of them counting 'one, two, three' – each makes one of three signs with his fingers:
clenched fist to indicate 'stone'
first two fingers outstretched for 'scissors'
flat of hand for 'paper'.

Aim. To win points recorded as beans, buttons or pebbles.

Rules. 'Scissors' beats 'paper' because they can cut it.
'Paper' beats 'stone' because it can wrap it up.
'Stone' beats 'scissors' because it blunts them.

The rules can be summarized in symbolic form: a picture that suggests a dog chasing its own tail.

Mathematicians use such kinds of arrow-pictures as an aid to their thinking.

Play. When both players present the same hand sign, neither wins. Then the tiddly-winks or counters are used as a decider: each player extracts a tiddly-wink from a bag in which there is one green and one red tiddlywink. Red beats green. After a draw, the player who draws red wins the round. The grab bag is actually a decision mechanism: it selects an order. The tiddlywinks represent a second order relation: red beats green.

Keeping the score. Use counters on a Ludo board.

Each time a player wins, he moves his counter forward a square, exactly as for 'Ludo'. Then the players can see their progress.

SET 8 **Time Games**

Telling the time. Time is a mystery to child and adult alike – but in different ways. Many mothers and infant teachers must have had the following kind of experience with four-year-olds: lunch is over. Mum is washing up. Her child comes in with yet another question; but this one is about time: 'Mummy! Is it tea-time?' Mum explains patiently that lunch is only just finished so it can't be tea yet. Usually the explanation falls on deaf ears.

The fact is, youngsters haven't the slightest notion of what time is. Philosophical considerations apart, a youngster does not sense – let alone understand – that time 'passes' as we say (and think we understand). Now without a feeling for the passage of time, it is hard to see how a child can possibly grasp what we mean by telling the time. Even by the age of six, when he may be able to read the clock, he may barely have the haziest idea of 'time'. Ask a six-year-old what *the* time is and he may be able to tell you. Ask him what a clock does and he may answer pat: 'It measures time.' But ask him what time is and you will be met with dumbfounded silence – quite rightly perhaps!

At the outset a child's conception of time hinges on the ideas 'before' and 'after'. The reader will recognize these terms as ordering relations, like those in Set 7. This set of games uses these time relations in the form of card games: they lead on to the relations of 'immediately before' and 'immediately after'. Not only are the games novel: they may, we believe, provide stepping stones across that Rubicon in a child's life known as 'telling the time'.

PASSTIME CARDS

4 to 5 years

A pack of eight cards. The games described below are intended to focus a child's attention on the general relations 'before' and 'after' and the particular relations 'immediately before' and 'immediately after'.

Things needed. 8 large postcards
Colored felt pens or ball-point pens.

Trace or copy the eight cards in the picture overleaf. Free-hand versions of your own are, as usual, preferable.

All the following card games are played much like 'Snap' for which you'll need two cards of each picture shown overleaf.

Snap

4 years

Play. Like 'Snap'. Deal out the cards evenly. Each of the players holds his cards face down, and turns over *together* one card at a time. They shout 'SNAP' if two cards are the same – the scenes happen at the same time. The first to say 'SNAP', but only when it is right to do so, wins the other player's snapped card and all cards under it, already played. The winner is the player who collects all the cards.

zzzzZZZ

An alternative version is 'NOT-SNAP' where the players call if the scenes are *not* the same.

'After' Grab

4+ years

Play and deal. The same as for 'Snap'.

All 'Grab' games are quiet games for two players.
Aim. To be first to spot the card that shows a scene taking place *later* in the day – that is, *any time after* the other card. The player who does has to point to the 'later' card and grab a cork (or ball of paper, potato or thimble). The first to get it right, wins all the opponent's cards so far played. Should the cards show the same scene, neither is 'later'. A player who grabs the cork *then*, loses his pile of played cards to his opponent.

'Before' Grab

The same game – except that the winning card shows a scene that comes *any time before* the other card.

Noisy versions of both games involve shouting 'Grab' instead of grabbing a cork.

'Next After' Grab

5+ years

Basically, the same game – the winning card shows a scene that comes *the very next after* (or immediately following) the other card.

Sample play. Bob plays a card showing a picture of getting up and Sue a card of going to school. Neither can grab the cork (or call 'Grab') because 'school' is not the next card after 'getting up'. Bob then turns up 'cleaning teeth' and Sue plays 'bed-time reading' (which *is* next in time order). Bob makes a 'grab' (correctly) but he mistakenly points to 'cleaning teeth'. Sue quickly points to the other card. She wins all the cards Bob has played.

'Next Before' Grab

The same game as 'Next after' except the winning card shows the scene *just before* (or immediately before) the other card.

My Favourite

4 to 5 years

A putting-things-in-order game.

Find the 'Happy Sets' Cards. Discard the 'small' pictures. Suggest your child puts the cards in some order. He can choose what he likes best first – all the dinosaurs, say; then next all the cars, then all the dolls, and last all the bears. Ask him to arrange the cards in that order:

dinosaurs
cars
dolls
bears

Suggest he tidy up the row by putting the colors in some order:

red
grey
black

The cards should finally look like the picture.

Next suggest he fans the cards, still keeping to his order, round in a circle – like a dog chasing its tail. See the picture.

Lead him on to see the following clock-face patterns. In our picture,

the bears extend from 12 o'clock to 2 o'clock,
the dinosaurs from 3 o'clock to 5 o'clock,
the cars from 6 o'clock to 8 o'clock
and the dolls from 9 o'clock to 11 o'clock.

The reader might have been tempted to have the dinosaurs begin at 12 o'clock because we listed them first. What matters is not *where* on the clock face the cards are, but their order among themselves.

Picture Clock Face

Things needed. Old kitchen clock (expendable!)

colored, gummed shapes – squares
hearts
circles

in these colors: red
black
pink
grey

It is not, of course, essential to have the shapes and colors listed here – as long as there are three shapes and four different colors. Alternatively, you can have four shapes in three colors – just like the 'Happy Sets' cards.

Take the glass front off the clock. Remove the hour hand or snip it off. Stick on colored shapes over the hours. Stick 'red square' over 12 o'clock. This is 'red square' time or 'red diamond' time, depending on how you gum on the shape. Stick on all the other signs round the clock face as in the picture.* Now put the glass back on. Wind up the clock and set it going. The minute hand only will move.

Play. Ask your child how long it will take him to play a game or do something special. He is not telling the time in the sense of knowing whether it is 4 o'clock: he's getting the feel of the *passing* of time. He might

* The double-order is here a little harder to follow but it fits together quite logically. You build it up this way: not only must you keep to the order of the things, as before; you must also keep to the color order *and* never have two of the same shapes next to each other nor two of the same color next to each other. Puzzle it out and you must hit on a pattern rather like our picture.

guess that it will take till 'grey square' time – in conventional time, a quarter of an hour. Evidently, a quarter of an hour passes between nearest different colors of the same shape. Encourage your child to see the other quarters of an hour on the face; they needn't be on the upright 'cross' at the usual 12, 3, 6 and 9 o'clock times. For instance, a quarter of an hour passes between 'grey heart' time and 'pink heart' time.

Look at 20-minute intervals – between 'red circle' time and 'red square' time, and between 'black heart' and 'black circle' time. So 20 minutes is the lapse between nearest shapes of the same color.

Leave the clock in the nursery or bedroom where your child can always see it.

SET 9 **Number Games**

The problem of number hang-ups. 'Please, Miss, is this an "add" or a "take away"?' is a question all too often asked. It reveals a complete lack of understanding by the child of the way to use his number skills. The misunderstanding is usually the result of his learning maths by rote. Agreed, he may have learnt to add and subtract with monotonous accuracy. And make no mistake about it, without such a basic skill he will never be any good at maths. On this count alone the mechanical rote method has much to commend it – though, it should be borne in mind, it is not the only possible method of learning arithmetic. As far as exams are concerned, rote learning can take a child a long way. It is only when it comes to problem solving that it lands him in difficulties. In musical terms, it is as if he knew the notes but not the tune. This is because his grasp of – and interest (such as it is) in – numbers ends at the grubby symbol so painstakingly scrawled in his exercise book. Of course, he knows it ought to be neatly written. But he also knows however tidy his 'sums', his mind will still remain cluttered with meaningless squiggles and signs. And that's about as deep as the rote-taught child's mathematical know-how is likely to go.

So on the one hand there's the nicely juggled symbol on the page and on the other, the ably handled image in the mind. How are we to bridge the gap? Our suggestion is – and there is scientific evidence to support it – to encourage a child *to invent his own symbols* especially in the early stages: if he does, he should be more able to solve problems; and he should do routine 'sums' all the better for it, and even with enjoyment. It's best, we find, deliberately to encourage children to chop and change their symbols rather than tie them to one tried-and-tested symbol – the one favoured by adults. This finding is after all no more surprising than the discovery that some children learn to read better on alphabets specially designed for them, like the initial teaching alphabet (i.t.a.). As with i.t.a., the child later discards his invented symbols for the orthodox ones. A child who invents his own symbolism or uses many different symbols, is likely to have a clearer picture in his mind of how to translate a problem on a page into symbols he can do math with. The reasons for these findings need not concern us here.

What concerns us are children's hang-ups over arithmetic. To help avoid them we have specifically created these number games: they bridge the gap between handling solid objects – counters, blocks and so on – and writing symbols.

An example may help to make this point clearer. A stick of five plastic cubes can be matched up, one for one, with the fingers of one hand. The stick itself can then be used as a handy symbol for 'five': its 'fiveness' is obvious from the number of cubes it is composed of. But the orthodox mathematical symbol isn't like that: there is nothing about the symbol '5' to suggest the number of fingers on a hand. Any more than there is about the ۵ which is the Pakistani '5'. Children are usually asked to operate with symbols without being allowed enough practice doing sums, first using solid objects and later solid symbols. Weaning a child off sums 'in the round' too early might be likened to asking a child from the Sahara to reason about the properties of snow from what he has seen of it in Dickensian Christmas cards.

Two big ideas. How we write numbers was a great invention. But there is another big idea in arithmetic – grouping. We write numbers right-to-left to indicate how we group things we have counted.

In ordinary counting we start with single things then group them in tens; then we group the tens into hundreds (ten lots of ten); and then group every ten hundreds into a thousand; and so on. When we write a number, such as 439, it is understood that the 9 refers to singles, the 3 to tens and the 4 to hundreds. In the child's eyes, and the mathematician's, the way we write numbers is purely a matter of taste. We could just as well write the same number, using a left-to-right convention, as 934. It is no more tiresome than changing from driving on one side of the road to the other to write numbers this other way round. It is simply a rule the child must learn. But the way we write numbers doesn't actually show the grouping that went on before we wrote the symbols – no more than the symbol actually resembles the number of objects it represents, as we said before.

Children need plenty of experience in grouping real things; and grouping the things in lots of different ways. They don't have to stick with grouping by tens, as we normally do. There are two good reasons for this. First, a child needs as wide an experience of number work as possible. Otherwise when asked to add two pears and three pears, he may reply that he has only learnt to add apples!

Second, we want a child to realize that ten isn't the only possible grouping number. For instance, we can group numbers in threes (as in the 'Sesame' game) or in fours (as in 'Arapanza'). The fact that computers group or count in twos – they reputedly use so-called *binary* numbers – is actually neither here nor there to a child! Though it may be a reason for an adult programmer relearning how to count. If a child can group in tens and threes and fours, say, then he stands a fair chance of understanding the basic idea of grouping. To show what we mean, suppose you ask a child to tell you the difference between 23 and 32. Many children will write the digits 2 and 3 down in either order when writing the result of a sum. But a meaningful reply might be: 23 means two groups of ten and three singles; whereas 32 is three groups of ten and two singles.

The following games can help to put meaning into a child's number work.

Sesame

6+ years

A counting game.

Usually we count in tens: in this game we count in threes – it is less to handle and easier to count out.

Things needed. Sweets, nuts, beans, counters or whatever for counting
Saucers
Table mats.

Aim. To show a child exactly how things are counted. A valuable spin-off may be that the parent or teacher is thrown into the deep-end as well as the child. He feels, with the child, what it is like to learn counting from scratch.

In ordinary counting the numbers go up in tens – from singles to tens to hundreds to thousands and so on. In 'Sesame' counting the numbers go up in threes – singles, threes, nines, twenty-sevens, and so on. To see what we mean glance at this example of grouping – first by tens then by threes.

Spread 23 sweets, say, laid out on a table. They don't have to be neatly arranged; in fact, for a child, the more higgledy-piggledy the better.

First, group the sweets in tens:

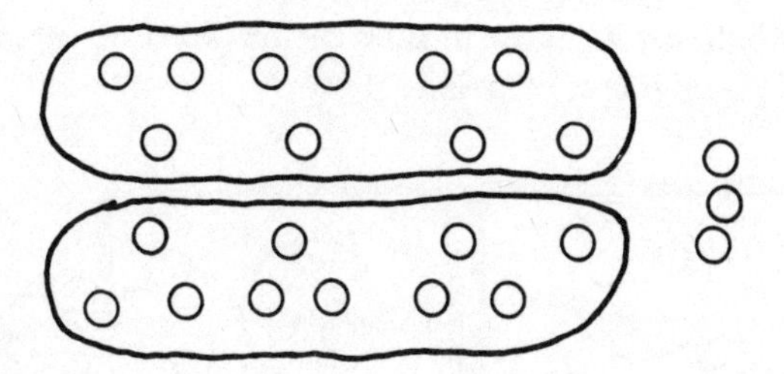

making two lots of ten and three single sweets. We can't group hundreds of sweets because it would take too long and we'd lose count! Still less thousands of sweets. This is where 'Sesame' comes to the rescue. Let's count the 23 sweets again, but in 'Sesame' counting as it might go with a child.

Count and group in 'threes'. It is essential for the child to touch each sweet as he counts. Otherwise he may miss one out or count one twice. The simple rule is, *touch and count.* As he touches the three sweets in turn, he can say the three syllables '*Se-sa-me*'. Alternatively he can say 'one–two–three'.

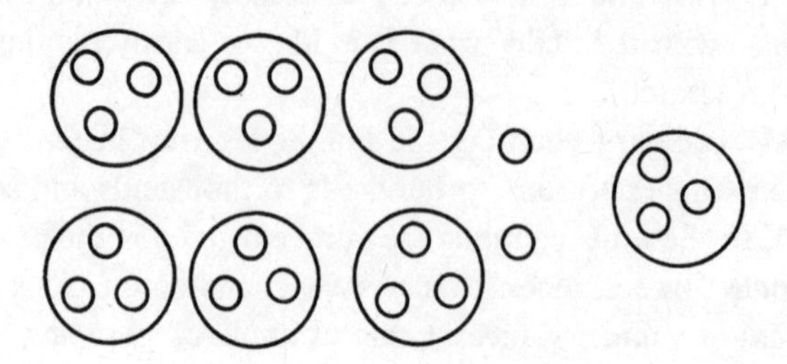

Each time he counts a group of three – a Sesame we'll call it – he puts the Sesame on a saucer, a 'Sesame saucer'. He groups the sweets into 7 Sesames and 2 sweets left over.

The next step is to group Sesame saucers in threes touching each saucer in tune to the word *Se–sa–me*, and placing three on a mat:

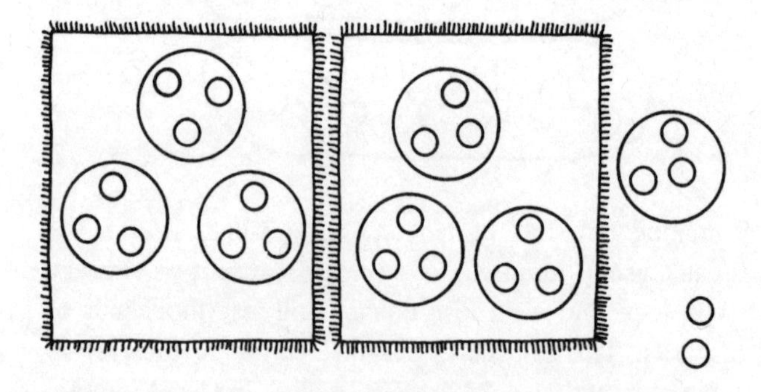

making two matfuls – call them 2 Big Sesames. There are no more threes of a kind to group together. So we have grouped as much as we can. We can *write* the total:

2 sweets, 2 Big Sesames, 1 Sesame.

Or you may be tempted to put them in order:

2 Big Sesames, 1 Sesame, 2 sweets.

Try various numbers of sweets up to 26. If you are feeling adventurous, try 27 or more. Then you will have to group three Big Sesames together to get one Giant Sesame.

Math note. But, you may be wondering, how is this remotely like the ordinary number 23? We don't usually write a number as 3 singles and 2 tens – yet they did once upon a time, as evidence the 'Four and Twenty blackbirds baked in a pie'. How can we leave out the words 'singles' and 'tens'?

The secret lies in putting the groups in order – highest on the left, lowest, the singles, on the right. So when we write 23 we always mean that the 3 on the right stands for 3 singles and the 2, next along, stands for 2 tens. And not the other way round.

But at the child's play-stage, all that matters is that he groups the sweets correctly. Certainly, never worry about writing the totals down – that is, symbolizing the totals – at this early stage.

Before your child gets tired of this game, move on to the next.

Arapanza

6+ years

Another counting game: counting in fours. The four-syllable title was suggested by one of the authors' daughter, Miranda.

Aim. Exactly the same as for 'Sesame', to help with counting and number. Where before we counted in groups of three, here we count in fours, on to saucers to the syllables of *A–ra–pan–za.* Let us see what the 23 sweets look like counted this way:

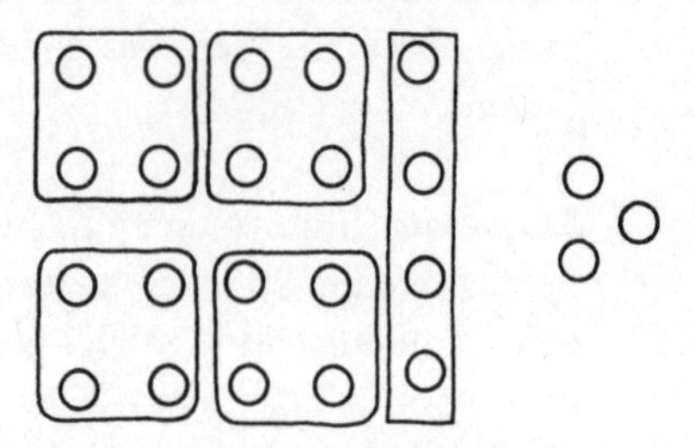

That's the first step done: 3 singles and 5 Arapanzas. The next step is to group the Arapanzas together, in fours, on mats:

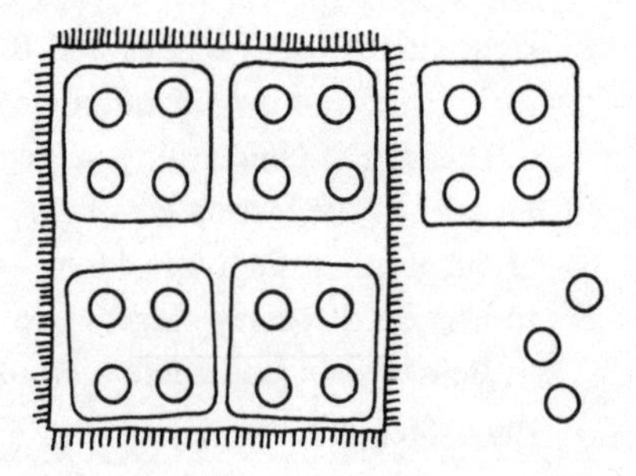

Which makes finally:

3 singles and 1 Big Arapanza and 1 Arapanza.

Again we have grouped as much as we can. Of course, the order in which you write the groups down depends on how they are placed on the page. Turn the page upside down and you'd get the reverse order! Anyway, your child won't yet be concerned with such high-falutin' detail.

Sesame Dice

6+ years

Counting game with a die in which any number can play.

Things needed. Large wooden cube (a child's brick)
Paint
Nuts, beans, counters, buttons or whatever for counting.

Aim. To collect nuts by throwing a specially marked die.

Mark the cube's faces like a real die but in this way (see the game 'Sesame'):

1 dot

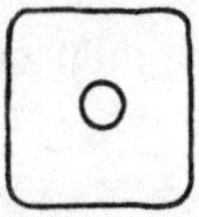

1 Sesame and 1 dot

2 dots

1 Sesame and 2 dots

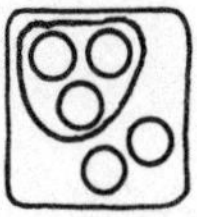

1 Sesame

2 Sesames

Play. Each player throws the die in turn. He picks up from a pool of nuts what is shown on his die. To make it more difficult, you can insist that each player call out how many dots are shown on the die. For example, suppose the highest face is thrown. The correct call is '2 Sesames' and not '6'.

After each throw each player must group his pile properly. If he doesn't, he loses a nut. The winner is the one who collects 2 Big Sesames first – that is, 6 Sesames.

Swaps and Shops

6+ years A game of handling money.

The problem. Although most children like playing shops, few completely understand what they are doing: they merely go through the motions of buying and selling. Neither a paper dollar bill nor pound note *look* more valuable than a metal penny. So a child cannot 'see' the value.

The purpose of this play situation is to help a child see the value of money from the symbols we use for it.

For instance, a stick of ten cubes patently has ten times as much stuff in it as an individual cube. The question is: How do we make paper money or metal coins give up their secret to the child, tell him what they are worth in terms he can understand?

A solution. One answer seems to be to convert the money counters into an equivalent value of stuff – plastic or wood blocks: they become solid symbols, rather than written ones, a child's cowrie shell money. The amount of wood or plastic considered equal in value to a coin is, needless to say, quite arbitrary. The blocks are nothing more nor less than a teaching device.

The child is now free to handle the blocks and assess for himself the coin's value from its equivalent amount of plastic or wood. He begins with coins, exchanges them for plastic blocks, exchanges big blocks for the going rate of smaller ones, and finally swaps back into smaller coins – smaller in value not in size! The more such shopping 'expeditions' he does, the more confident he will become in handling coins on their own. Eventually he will throw away the crutches of blocks – but only when he is good and ready.

The game. In this game, the child swaps 'pretend' coins (counters) for plastic or wood blocks, which become temporarily a look-and-feel monetary standard. The child can see and actually feel the value of his money from the amount of stuff he swapped it for.

The game begins when he goes to the Toy Shop to buy something. Make sure he always starts out with a large denomination coin so he will need to change it for smaller coins. He gets his change by what may appear to be an unnecessarily roundabout method. Actually, it is not so much roundabout as carefully planned: it is a series of step-by-step manoeuvres to give the child a mental picture of equivalent coinages.

To get his change, then, he must first go to the Bank. There he must change his coins into blocks – a blue coin for a big block and so forth. But he's not through yet. He must next proceed to the Swap Shop where he can 'break up' his blocks – or rather swap them for smaller pieces. It is in the Swap Shop that he can see the equivalence of stuff before his very eyes. Then he returns to the Bank to turn his newly swopped blocks into smaller denominations of coin. Only now can he go to the Toy Shop and buy the toy of his choice – provided he has the right change! If not, he must go through the whole rigmarole again. The description sounds longer than the action takes in practice.

The Toy Shop, Bank and Swap Shop can be set up in the play area as shown here.

Things needed. Brightly colored counters – yellow, red, green and blue (any four colors will do)
(*a*) 40 plastic cubes such as 'Unifix' cubes, or
(*b*) wood dowelling, 2 cm by 2 cm square cross-section.

(*a*) *Plastic blocks.* If you are going to make the play-withs of plastic blocks, buy ready-made blocks, such as 'Unifix'.

Make up several sticks each of three cubes press-fitted together.

Make up squares by binding three sticks together with Scotch tape or with rubber bands.

Make big blocks by binding together three squares.

(*b*) *Wood blocks.* If you decide to construct the play-withs out of wood, which is a lot neater but more troublesome, buy square-rod dowelling, 2 cm by 2 cm.

Chop up 20 or so little cubes out of the dowelling.

Cut up about 30 sticks, each 6 cm long – that is, equal to three cubes stuck end to end.

Make squares of wood by glueing three sticks side by side; or tape them or bind them with rubber bands; or you can cut them out of wood plank, 2 cm thick.

Make the big blocks by sticking three squares together. Or you can cut them out of square rod, 6 cm by 6 cm.

Whether you use method (*a*) or (*b*), or a mixture of both, you arrive at these playwiths:

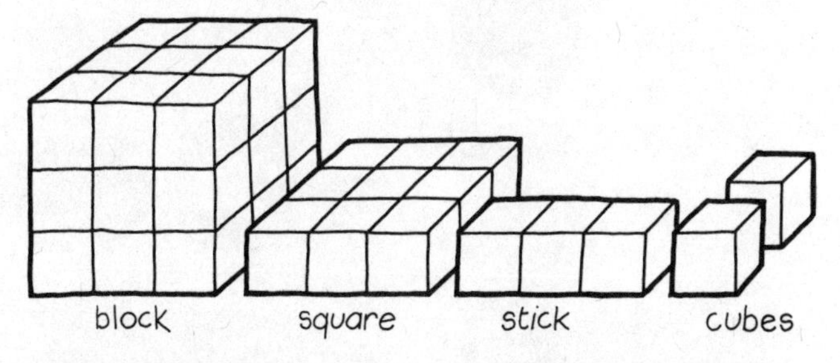

The Bank. Stock the Bank with black, grey, red and pink counters – the coins. The currency values are shown in the picture.

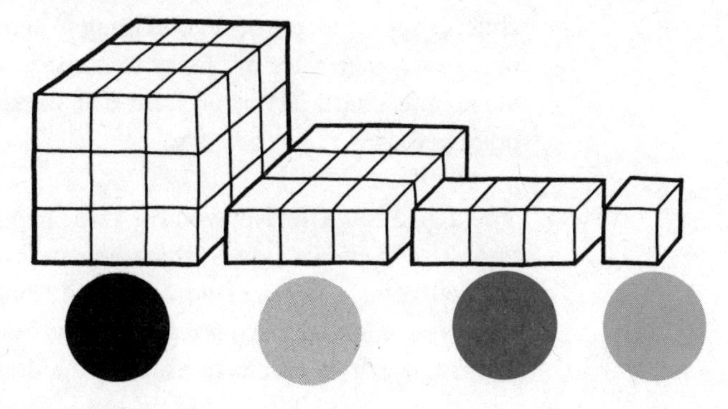

A black coin is worth three grey; a grey is worth three red; a red is worth three pink. The coins are grouped in threes – just like the 'Sesame' game. Make out a chart of the currency large enough for your child to see. Or fix spare coins to the chart.

The Swap Shop. A simple list shows the swapping rates for coins into plastic or wood blocks. The number of blocks tallies with the nominal value of the coins. The Swap Rate is shown in the picture.

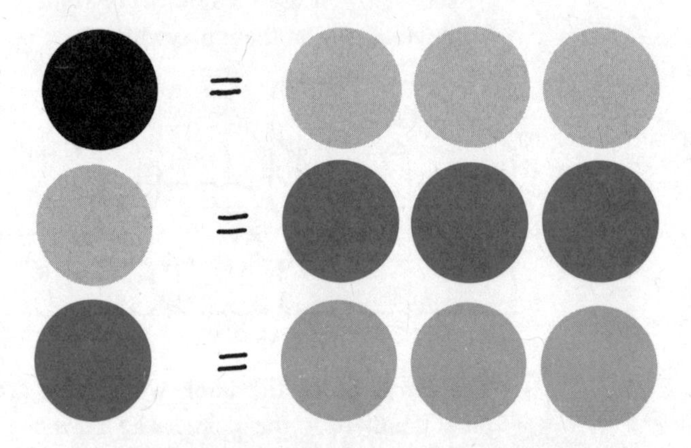

Again, this should be displayed large and clear for your child to see. It might be a good thing to show the combined exchange rates of blocks to coins to other coins all in one chart. Try it and see if it helps to get the point across.

The Toy Shop. On display in the Toy Shop are various toys – your child's own – which he can 'buy' with the colored coins. There is no reason, if things go well, why you shouldn't slip in a specially bought cheap present which he can have provided he 'buys' according to the rules! The aim then would be to make sure he has to exchange his coins via the Swap Shop. Our picture shows some suggested prices. If you give your

child one black coin or two black coins he is bound to need to go through the Swap Shop to buy the toys at these prices.

Play. Play revolves round the three centres, the Bank, the Swap Shop and the Toy Shop, set up in different parts of a room, shown earlier.

Your child begins with, say, one red coin in hand. He goes to the Toy Shop where he wants to buy a toy animal for two pink coins. He needs to change the red coin to be able to buy in this game. So he goes to the Swap Shop and with his red coin buys a stick of wood (see Price List). But he needs to break this down even smaller. The Swapper (you) willingly gives him three small cubes in exchange for the stick. The child takes the three cubes to the Bank which pay him three pink coins – one for each cube.

The child then returns to the Toy Shop and buys one toy animal for two pinks, leaving him one pink in hand.

The game may seem lengthy and unnecessarily complicated. But, when played, it is really quite simple. It shows how we exchange money in terms a child can understand.

Problem item to try. The reader may prefer to glance through the strip cartoon of the same sort of situation (see the picture on the right).

In hand: 1 black
To buy: a fire engine (1 grey, 1 red).

Clumsy solution. Turn the black coin into one big block of wood. Then into little cubes and back into 27 pink coins. While this solves all exchange problems, it is a bit like paying for one's shopping all in pennies. In other words, it is best to deal in as large value coins as possible.

Neat solution. At the Bank turn the black into one big block and at the Swap Shop swap *that* for three squares. Change one of the squares only into three sticks. Back at the Bank cash the two squares for two grey coins and the three sticks for three red coins.

To the Toy Shop, where the child can pay for the fire engine with one grey and one red coin, leaving in hand one grey and two red coins.

Mr Aladdin

4 to 5 years A multiplying ('times') game.

Things needed. Lots of little blocks, sticks and counters
Farm animals
Dolls' tea sets
Toy cars
Trains and so on
Large sheet of paper or blackboard slate.

1 Sam has one black coin
2 At the toy shop
fire engine
price
3
that's no good --must go to the bank
4 At the bank
one block Sam for your black coin
5 later at the swap shop
Sam has his block cut into three squares
6
and chop one into sticks please
7 then Sam goes back to the bank to exchange wood for blocks
thats 2 grey and 3 red
2 squares and 3 sticks
8 Sam races to the toy shop
1 grey and 2 red
thats right, here's your injin Sam
got me injin and got these left

Aim. An adult impersonates Mr Aladdin who offers so many toys, from the nursery, for every 'coin' (block) the child wants to spend in Aladdin's cave. Aladdin might offer, say, three toys for every block brought by the child. He could call his 'price' '3 for 1'. (In effect, then, Aladdin acts as a multiplier.)

Play. An example will show how the game goes.

Mr Aladdin has a magic table in a cave. On it he has his wares – the nursery toys.

THE CHILD: Hullo, Mr Aladdin! What's your price today?

ALADDIN: Wait and see . . .

He writes his price on the board, 4 for 1. He writes this as:

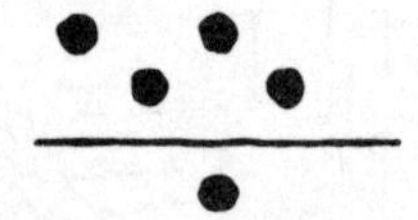

that is, four dots over one dot which is intentionally similar to the fraction or ratio $\frac{4}{1}$. Always use dots and not numbers for youngsters. It is a temptation to use a square pattern. Resist it. A random, higgledy-piggledy pattern promotes learning (actually, abstraction).

ALADDIN: Four for one, four for one today. What would you like – animals or cars?

THE CHILD: Animals.

ALADDIN: Plenty of *them*. That's fine.

The child counts out three 'coins', say. (Too large a number may exhaust the stock of farm animals!) The child places the 'coins' on the magic table one at a time. For each coin Aladdin counts out his 'price' –

here, four animals for each coin – that is, 12 animals in all.

Then the adult and the child swap roles.

Other prices Aladdin could try:

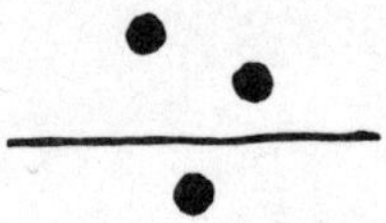

= 2 for 1

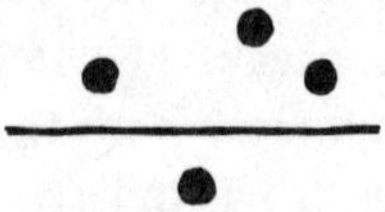

= 3 for 1

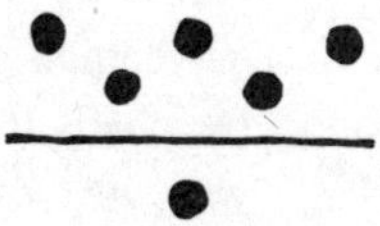

= 5 for 1

More interesting prices are:

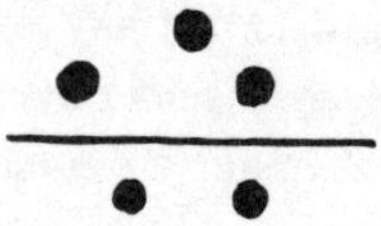

= 3 for 2

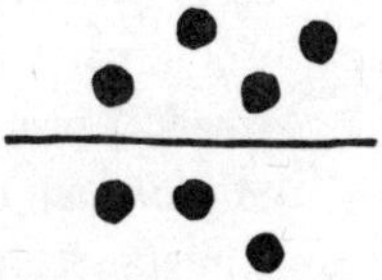

= 4 for 3.

Suppose Aladdin runs out of a particular toy. Then Aladdin can but offer nothing (zero) for 1, for he has no animals to sell. He writes his price, then, as

$$= \frac{0}{1}$$

This is the child's first game about zero.

SET 10 **Pebble Games**

Symbol games. All too often a child is faced with oceans and oceans of chalky symbols on the blackboard, which mean nothing to him. To keep his head above water he has to learn them by heart. The Pebble games have been devised to help youngsters see what lies behind math symbols. Where the Number Games were aids to handling numbers, the Pebble games help with very early algebra – that is, math with letters. In the games the child is not asked to pore over numbers and letters: instead he moves colored pebbles or buttons or hares about a field. So any resemblance between the games and algebra is not obvious but entirely intentional! For a mathematician *could* if he wanted solve his equations by using pebbles instead of x's and y's – much as an author could use picture writing instead of printed letters. But it would be rather slow and impractical. However, unlike picture writing, the pebbles can stand for abstract relations you cannot really illustrate. Herein lies their power – and their relevance to maths.

Many of the games have the same underlying pattern. The cutting-out activities and the Rainbow Toy game in Sets 4 and 5 were amazingly, as we saw, merely disguises of the same pattern. Equally astonishingly, 'The Pebble Game' has the same form as 'Rabbit Hunter' in this Set of games. The one consists of moving pebbles like the Japanese game of 'Go', the other of haring about a field. Yet at root they are the same: the same *kind* of change takes place: it is simply that different *things* change in the two games. 'Rabbit Hunter' is but 'The Pebble Game' writ larger – and more energetic!

As usual, we ask you not to tell your child about the similarity between games. Let him come to it for himself. Anyway, he may not be ready to grasp what you are saying. In which case he will merely mouth your words back at you. In telling him, you may also spoil the fun.

The Pebble Games form a sound basis for learning 'clock arithmetic', now taught in primary school. They also convey a strong hint of what algebra is really about.

The Pebble Game

5 years and up; in fact most ages

Things needed. 10 black and 10 white pebbles, or buttons or counters (of two distinct colors).

Play. Cast at least six black and six white pebbles in a row so that they are well-mixed in order. (Later the children can do this.)

Aim. To shorten the row of pebbles as much as possible by removing or inserting pebbles in the row obeying the following rules. The winner is the last person to remove a pebble.

Rules. You may remove (or insert) anywhere in the row –

2 adjacent black pebbles
2 adjacent white pebbles
or a black pebble from each side of a white pebble.

Present these rules to a child with the pebbles laid out as a picture:

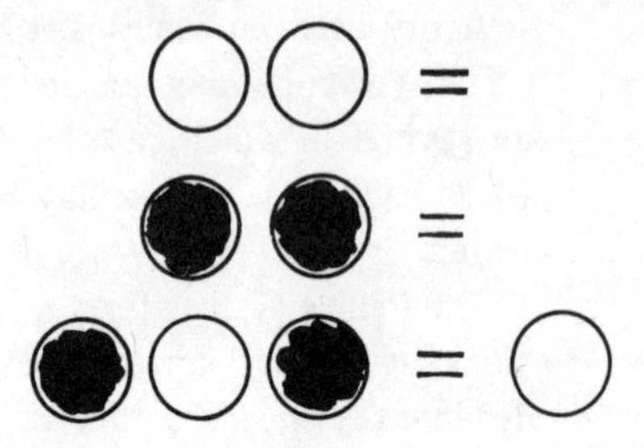

Example of play

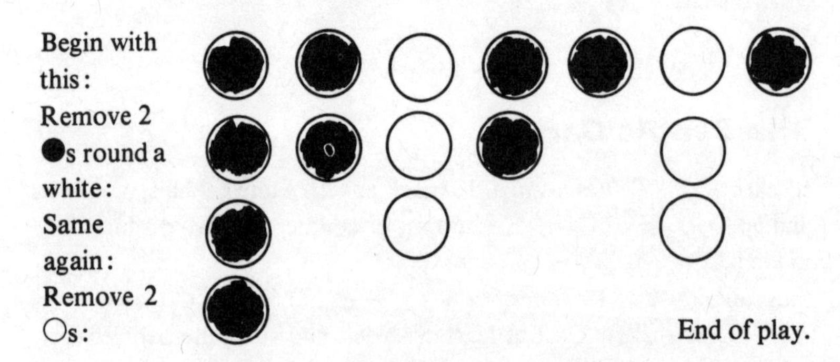

A row of seven pebbles was shortened to one black pebble, which cannot be removed. So that's the end of the game. The winner would have been the player who removed the last pebbles – the two white ones. The game stops when any of the five endings shown in the picture are reached.

The five endings

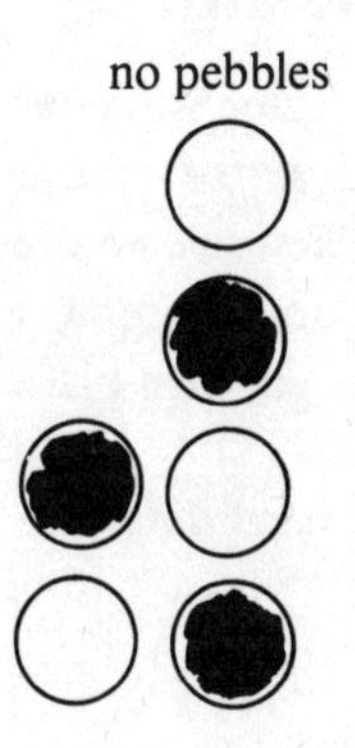

An interesting point about this game is this:

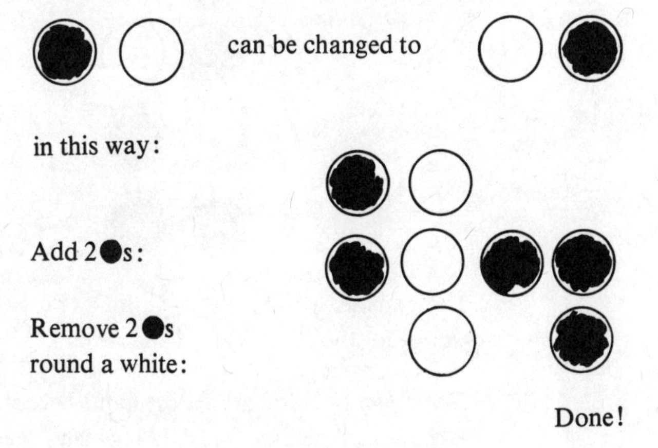

This means, paradoxically, the order of next-door pebbles can be changed without affecting the outcome of the game.

Black and White

6 years and up; in fact most ages

The game is the same as 'The Pebble Game' except for the first rule, and, as a result, the possible endings. The first rule now reads:

You can remove (or insert) anywhere in the row three adjacent black pebbles.

The six endings

no pebbles

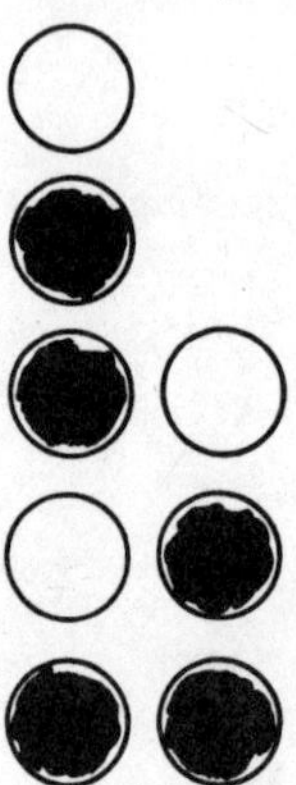

The endings are the same as for 'The Pebble Game' except for the additional one at the end.

Game blocked. Sometimes the game becomes 'blocked' even though one of these endings isn't reached. When the children are beginners, treat the blocked end as a proper ending. When they get more expert show them how to unblock it by inserting more pebbles.

Example of a 'block':

Blocked!

You cannot shorten it yet it isn't an ending. So try lengthening with 3 ●s on the far end:

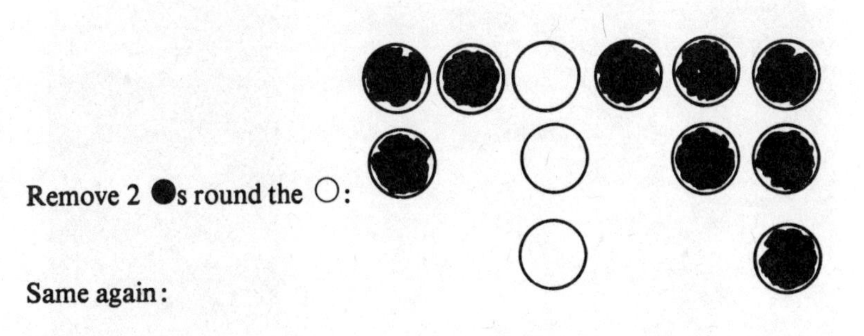

End of play.

Rabbit Hunter

4+ years

The game contains a big new idea!

A game for two players, plus an adult caller.

The game is played in a large oblong space, longer than it is wide – a garden, a yard, or a hall. An oblong could be marked out with clothes in a park, or scratched in sand on a beach or in snow in a field. The players begin by standing each at a corner. They move along the edges of the oblong in response to the adult caller's calls.

Our photograph (overleaf) shows an indoor version of the same game using a cardboard flipper, which some older children can be seen holding to take the place of the moves round the field.

Aim. For one player, the 'hunter', to catch the other, the 'rabbit'. The rabbit has two moves, the hunter only one.

The moves. The adult calls out moves for each child to make:
'along' move along the long side of the oblong

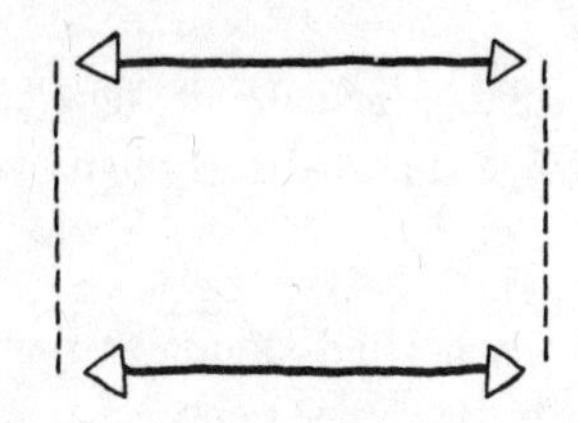

'side' move along the short side

'across' dance – along the diagonal.

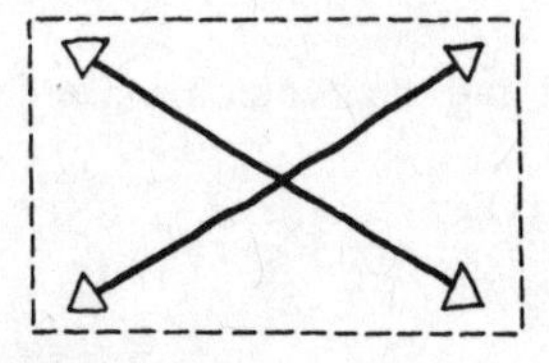

The big new idea. When the call is 'along', a child must move along either of the long sides and in either direction – depending on where he starts from (shown below by the circle):

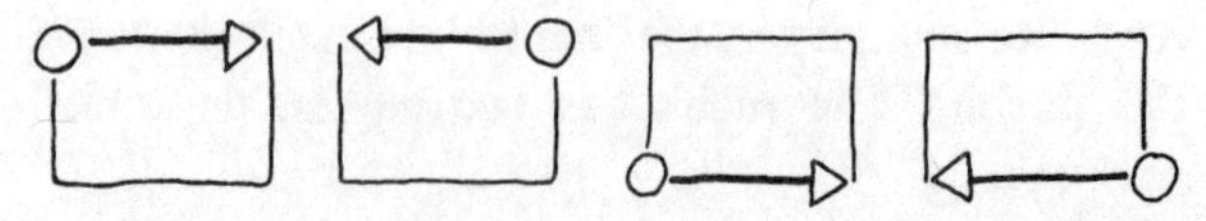

And the same holds for the 'side' and the 'across' moves. This means the moves do not have one fixed path. What they have is a definite two-way *direction.* This is a key idea in math. (It underscores the concept of a *vector.*)

Give each child, especially if very young, several practice calls to get him used to the moves and especially to the key idea.

Play. To start, each child stands at the same corner. One is the rabbit, the other the hunter. Call two moves – say, 'along' and 'across'.

The 'rabbit' must move first 'along' then 'across', as shown. The second move always starts where the other stopped.

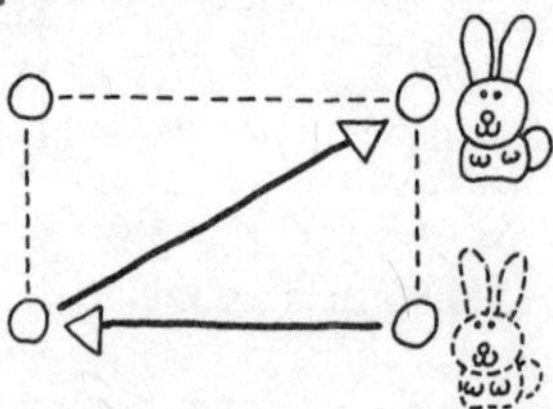

Now the rabbit and the hunter are like this:

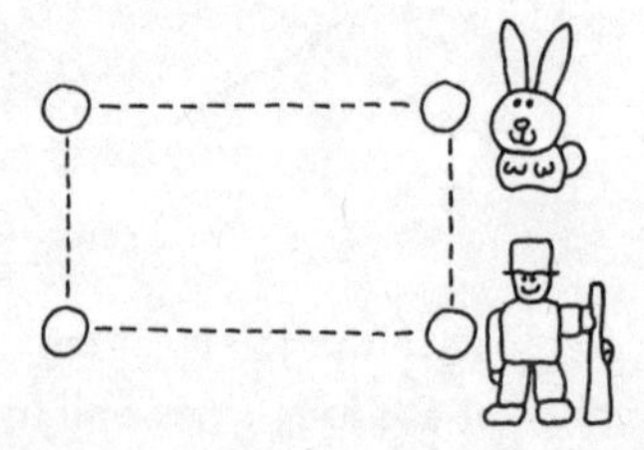

The hunter has to make the correct *single* move to catch the rabbit. It is clearly a 'side' move.

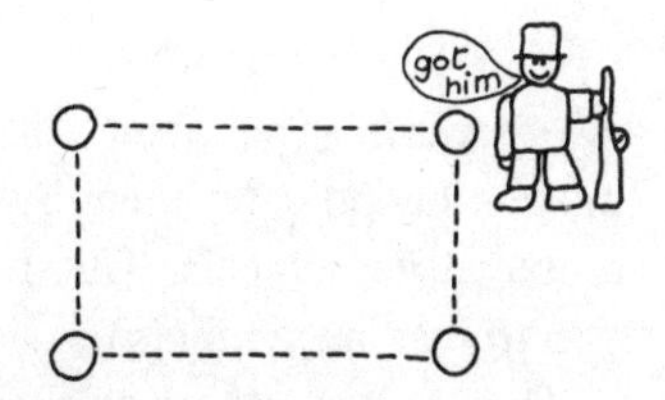

Some other double moves are:

'across' + 'side' = 'along'

'side' + 'along' = 'across'.

What happens if the caller calls 'across' and 'across'. Then the rabbit is back where he started, at the hunter's feet. What move must the hunter make to catch the rabbit now? No move, of course! The hunter can show the 'no move' by jumping up and down on the spot.

After five turns, the children can swap roles.

Alternative roles they can play – although the game is precisely the same – are:

cat and mouse
farmer and pig
tiger and goat
giant and Jack
witch and child

all to be acted out with noises and pantomime. For children of six years or older, the game could be played on a board with counters. But it would not be nearly so effective as the children themselves moving.

Pirate Flags

6+ years

This game is a follow-on from the last game. For four or fewer players. We'll suppose four are playing.

Things needed. Colored hankies or the pebbles.

The moves. Exactly the same as for 'Rabbit Hunter'.

The signals. The adult caller does not call out the moves: he indicates them by signals – like pirate flags or naval flag signals. This provides a gentle introduction to reading symbols for a child.

The caller waves colored (red, say) and white hankies to signal the moves –

white hankie for the 'along' move
red hankie for the 'side' move.

The caller lays the hankies out in a row, like a naval flag signal. Each of the four children stands at a corner of the oblong. Try some practice moves first, simply

calling out the moves aloud. The call is 'along' and, as in the picture, Ben and Ann swap places and Jill and Sam swap over:

The 'side' move looks like this:

The 'across' move with its red 'n white signal looks like this:

Play. Once the moves have been understood, play can commence. The caller 'runs up' this signal:

white hankie and red hankie

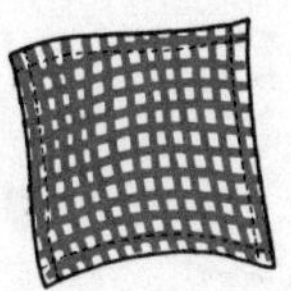
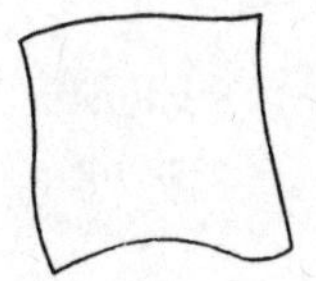

which signals 'side' move and 'along' move, the same as 'across' move.

Note. There are only two signals although there are three different moves. This makes the children *think.*

Call 'Ready, Steady, Go!' and everybody must rush to their new places as quickly as possible – or the pirates will get them! The last child to get there is 'out'; he sits out until the next game. Flash other signals (two hankies is quite enough for beginners: even mix in some single hankie signals) until only one person is left in. He is the winner.

Points for thinking. Award points – counters, beans or whatever – for 'thinking'. The winner of the race can win an extra point by indicating the signal for the move that gets him back to where he started from.

An example. Ben wins this race where the signals were white hankie then red hankie which signalled 'along' move then 'side' move = 'across' move.

Ben's moves were:

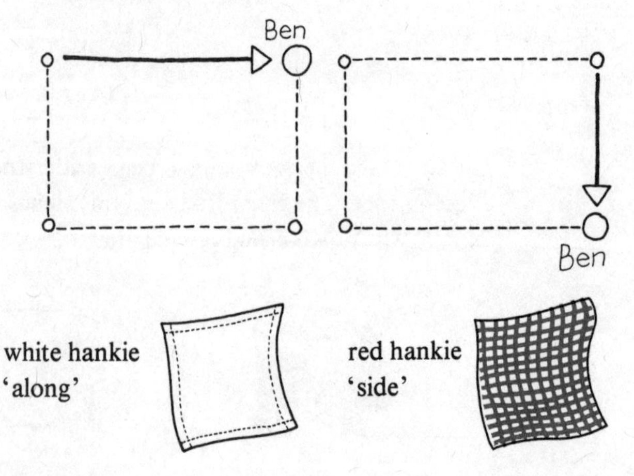

What single move gets him back to where he started from? The 'across' move. Which means he must pick up the red *and* the white hankies together – the only way of signalling the single move.

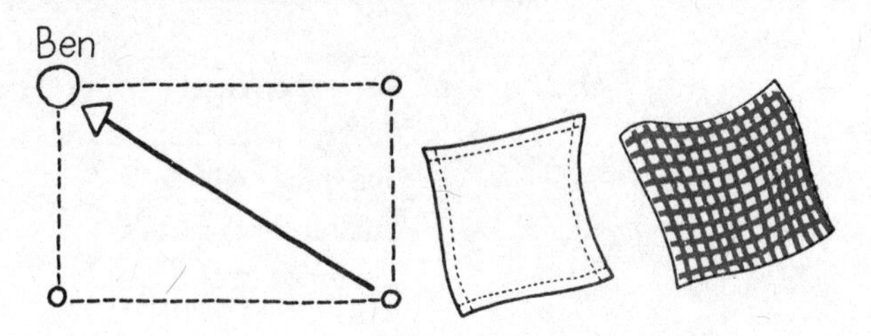

Space Race

7+ years

An older, 'thinking' version.

Play. Like 'Pirate Flags' except that the caller calls out *two* moves in succession: he does not use flag signals.

Aim. For the players to make one move only in order to win the race. This means thinking out the single move.

When two moves ('across'+'across', say) are the same as no move, the players hop to indicate 'no move'.

Example. The caller calls 'side and side'. Nobody need move – they hop up and down. Suppose a player *does* make two side moves in succession: he will find himself back where he started . . . meanwhile he will have lost the race!

Think or Run

7 years and up

A follow-on from the last game, it requires thinking skill as well as speed. The caller now signals (not calls) a succession of several moves.

Things needed. Several colored (red, say) hankies
Several white hankies but other distinctive things will do –
Buckets and spades
Tin mugs in two colors.

Play. As in 'Space Race', the thinking bit is a vital part of the game. The caller signals several hankies.

Aim. For the players to make the shortest number of moves to finish the race first.

Example. The caller displays this signal:

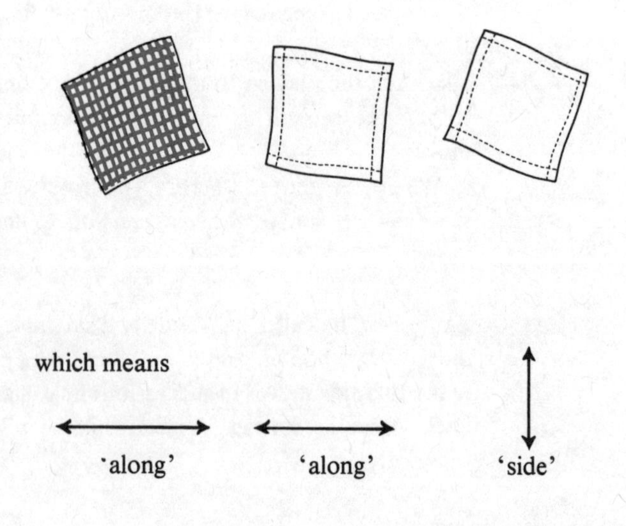

Jill reads the signals correctly:

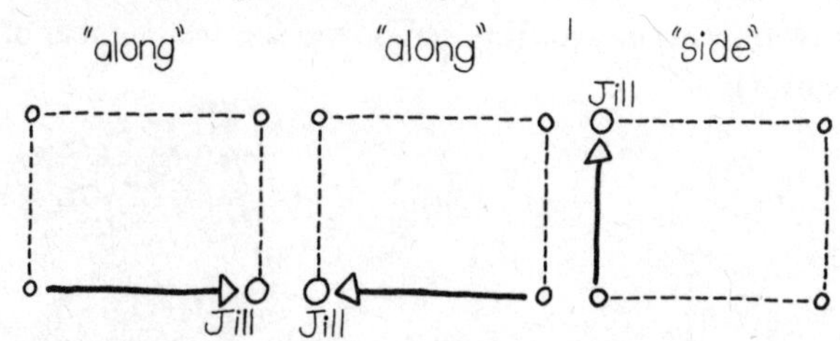

But Ann does some 'thinking': she figures quite correctly that 'white' then 'white' hankie in effect signals 'no move'. So she only has to make a side move. If the others race two lengths and a side of the oblong, she wins easily.

Armchair Race

10 years and upwards

Here's an armchair method, designed with Dad in mind, to work out the signals. Dad flicks through a row of pebbles in comfort instead of running round the oblong.

Things needed. Black and white pebbles.

Play. Put the pebbles down in a row – a black pebble for every red hankie signalled and a white pebble for every white hankie in the same order as the signal. So a message and its row of pebbles read:

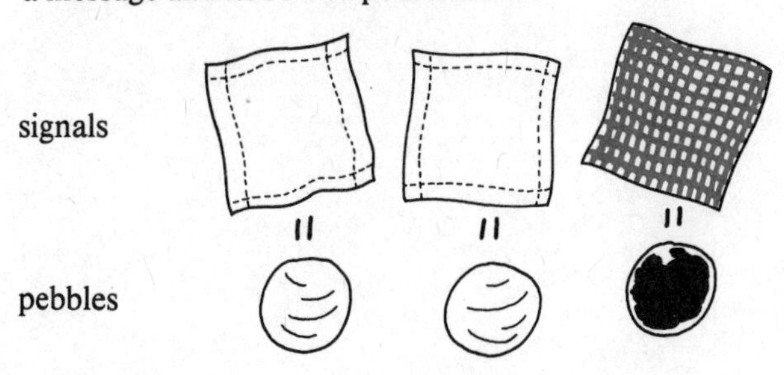

Now apply the rules of the *Pebble Game* to shorten the row (this exactly mirrors getting the shortest number of moves):

Remove 2  s:

End of play.

The remaining black pebble, translated back, means a red hankie which is the signal for a 'side' move. An astounding link holds between the hankie game and the pebble game yet they seem so very different. The fact is, both are based on the same mathematical pattern. (That's how we made the games up!)